THE BEGINNER'S GUIDE TO

Mushrooms

This book is dedicated to Gary Lincoff.
He was the greatest educator and biggest
inspiration to beginning mycophiles everywhere.

Brimming with creative inspiration, how-to projects, and useful information to enrich your everyday life, Quarto Knows is a favorite destination for those pursuing their interests and passions. Visit our site and dig deeper with our books into your area of interest: Quarto Creates, Quarto Cooks, Quarto Homes, Quarto Lives, Quarto Drives, Quarto Explores, Quarto Gifts, or Quarto Kids.

First Published in 2021 by Quarry Books, an imprint of The Quarto Group, 100 Cummings Center, Suite 265-D, Beverly, MA 01915, USA.
T (978) 282-9590 F (978) 283-2742 QuartoKnows.com

Quarry Books titles are also available at discount for retail, wholesale, promotional, and bulk purchase. For details, contact the Special Sales Manager by email at specialsales@quarto.com or by mail at The Quarto Group, Attn: Special Sales Manager, 100 Cummings Center, Suite 265-D, Beverly, MA 01915, USA.
10 9 8 7 6 5 4 3 2

ISBN: 978-1-63159-911-8

Digital edition published in 2021
eISBN: 978-1-63159-912-5

Library of Congress Cataloging-in-Publication Data is available

Design: Sylvia McArdle
Cover Image: (top) Britt A. Bunyard, (bottom) Shutterstock
Page Layout: Sylvia McArdle
Photography: Britt A. Bunyard

Printed in China

Disclaimer: This is not an all-inclusive resource for determining differences between mushrooms. Other references and experts should be sought out to help determine specific types of mushrooms. No mushroom should be eaten unless edibility is absolutely certain. Assume that all mushrooms are poisonous until proper identification is made. Even at that point, eat at your own risk!

THE BEGINNER'S GUIDE TO
Mushrooms

Britt A. Bunyard & Tavis Lynch

Everything
You Need to
Know, from
**Foraging to
Cultivating**

QUARRY

CONTENTS

PART I
Mushrooms: The Basics

READ THIS BOOK BEFORE YOU PICK YOUR NEXT MUSHROOM. This book is where your mushrooming journey begins, because this book is information. And with a little easy-to-follow information, the world of wild mushrooming will open up to you. This book will help you to identify the next mushroom you see in the field or forest, tell you a little bit about what that mushroom is doing there and its role in the natural world around you, and maybe even tell you some enjoyable things to do with it; for example, photography, natural fiber dyeing, cultivation at home, or putting it in a starring role in your next gourmet meal. The information in this book can also keep you safe, as many wild mushrooms are poisonous if eaten—some of them deadly. It is unwise and potentially dangerous to eat any mushroom (or plant or mystery meat at the highway truck stop, for that matter) without knowing exactly what it is that you are putting into your body.

ABOUT THIS BOOK

This book is a pictorial guide to the most commonly encountered mushrooms of North America and Europe. References to "West" means west of the Rocky Mountains, and "East" is east of the Rocky Mountains. In some cases, the geographic range will be noted as restricted to the Southeast, Northeast, Midwest, or the like in North America. For Europe, general geographic restrictions, if any, will be noted as "Mediterranean," "northern Europe," and so on. Besides geographic location and time of year, this book bases identification of mushrooms on macro-characteristics of mushrooms (those that can be seen with the unaided eye) as well as, at times, tastes and smells. It is absolutely safe to handle any mushroom—even deadly poisonous species. It is absolutely safe to taste a very small piece of any mushroom—even deadly poisonous species—as long as no part is swallowed or ingested. Poisonous mushrooms are dangerous only if consumed. Also note that mushrooms can at times be highly variable in appearance; size, color, and other features—even geographic region or season—may fall outside of what is described here or in any book.

Fungal hyphae

Fungi (the singular is *fungus*) are curious organisms grouped together by virtue of having a unique cell wall and physiology (it's chitin-based, as opposed to cellulose-based like plants; or cells without cell walls, as with animals, for example) and a unique form of reproduction and reproductive structures. The body of the fungus is called a **mycelium**; it is a network of filamentous cells that is the carrier of nutrients and water to a mushroom; each of these filaments is called a **hypha** (the plural is *hyphae*).

There are more than 100,000 named species of fungi on the planet, though mycologists (scientists who pursue the study of fungi) suggest there are more than a million (and likely several times that number) species awaiting discovery. Fungi have a heterotrophic lifestyle, which means, like animals, they cannot create their own energy and basic cellular materials from an energy source in the environment (in contrast to plants, which can harness sunlight energy). Therefore, most fungi make a living by doing one of three things in the environment: decomposing other organic matter that was once living, parasitizing a living host (many animal and probably most plant diseases are caused by fungi), or becoming beneficial partners with other organisms, especially plants. The more we learn about fungi, the more we see that the healthy function of the natural world around us is closely tied to them.

A subset of fungi produces reproductive structures that are large enough to be seen with the unaided eye (macrofungi, as opposed to tiny microfungi). We call those reproductive structures mushrooms. And to make things easier, we often refer to the fungi that make those macro structures as mushrooms too.

There are thousands of species of wild mushrooms in the northern hemisphere. Most have an important role in the environment. They grow in a wide variety of habitats, and most of the mushrooms seen on a walk in the woods are beneficial. Many species are quite specific about their food source and will be found under or near only certain kinds of trees. Some are important as decomposing organisms, aiding in the breakdown of logs, leaves, and other organic debris. This results in the recycling of essential nutrients. Mushrooms can also be parasites of trees or other plants. They can grow into and form their fruiting structures on living trees, thus causing decay of either the sapwood or the heartwood. Many woodland mushrooms, such as mycorrhizal mushrooms, are essential to the good growth and survival of trees because they establish a mutually beneficial relationship with living tree roots. All mushrooms—whether poisonous or edible—can be admired for their beauty and for the fantastic variety of their form, color, and texture.

Sometimes edible and poisonous mushrooms can look very similar, even growing in the same habitat and during the same season. Edible mushrooms are known to be safe to eat because they have been eaten frequently with no ill effects. Poisonous mushrooms are known to be dangerous because someone ate them and became ill or died. There is no test or single characteristic to distinguish edible from poisonous mushrooms. This indicates a need to identify *with certainty* one of several of the proven edible species; only after identifying a mushroom as edible should it be picked and/or eaten. There are no shortcuts to identifying wild mushrooms. The only way to know the edibility of a mushroom is to identify it with 100% certitude. Learn the edible mushrooms but learn the common poisonous mushrooms too. It is especially important to learn the characteristics of the *Amanita* genus, because several of its mushrooms are poisonous and common. *Amanita* mushrooms cause 90% or more of the mushroom poisoning deaths in the world every year.

Poisonous mushrooms do not have warning colors (or tastes or smells) like plants and animals do. Animals do not know when a mushroom is poisonous or edible. At the same time, some animals can eat mushrooms that we cannot.

There are dozens of myths about testing the edibility of a mushroom. None of these are based in fact, and all of them should be dismissed. Not all mushrooms growing on wood are edible, for example. We cannot assume that all white mushrooms are edible, and at the same time, we cannot assume they are all poisonous. Stick to accurate identification. It is the only reliable way to determine the safety of any given mushroom.

GUIDELINES FOR COLLECTING WILD MUSHROOMS

- Be sure of identification. Eat only mushrooms that are known to be edible.
- Do not eat raw mushrooms.
- Eat only young, fresh mushrooms.
- Eat only one kind at a time to isolate any troublesome species. Mushrooms can affect different people differently.
- Do not gorge yourself on wild mushrooms or consume them for many days in a row.
- Eat only a small amount the first time you try a new mushroom; even morels, generally considered to be excellent, can cause illness.
- Obtain a copy of one or more books or publications on mushrooms and cross-reference information.
- Get help from a local expert or join a mushroom club.
- Do not force your sense of adventure on others.
- Do not drink alcohol when trying a new mushroom species. Some species can be problematic with alcohol.
- Collect ethically. Leave the forest cleaner than you found it; carry out all your trash

plus any that others have left behind. Try not to take all the mushrooms; leave some for others. If someone is already picking on a patch, do not dart in and jump their claim.

- Observe laws for collecting mushrooms. Never pick on private property without first asking permission.
- If you are with a group, stay in close proximity to the group; never be the last one back to the cars or make others wait for you unless you're prepared to share a horde.
- If you are on an organized foray, find out if it's okay before bringing your dog—it's usually not. It may even be against the law in the forest you're going to. Be mindful that dogs could disrupt someone's photography setup and that some people are fearful of dogs.
- Share your knowledge and enthusiasm.

Parts of a Typical Mushroom

The lethal *Amanita phalloides* is one of the most infamous mushrooms in the world.

In order to navigate a book about mushrooms or to describe your mushroom to someone else, you will need to learn a few technical terms for the parts of a mushroom.

The **cap** is usually the first feature to study when you are trying to identify a mushroom fruit body. The cap (or pileus) is the top part of the mushroom. When immature, the cap can be round, conical, bell-shaped, or convex. As the mushroom matures, the cap flattens out. Other caps can become vase-shaped or even have a knob on the very top. Mushroom caps can be dry, wet, or sticky. There are also many textures to caps that help to differentiate one from another.

When a mushroom is turned over, either **gills**, **pores**, or **spines** will be seen. Gills appear as long openings from the edge of the cap to the stalk and are separated by thin tissues. Gill attachment to the stalk plays a large role in identification. Gills can be notched (or adnexed), free (not attached to stalk), or completely attached at a 90-degree angle. Some gills can run down the stalks. They can also be fringed or toothed, or they can have various colors. Spacing of the gills is also important.

Pores, on the other hand, are like long tubes running through the underside of the fruiting body. They appear as small holes. These holes are where the spores are released.

Spines hang from the underside of the cap. The spores are formed on the outside of the spines. In common usage, mushrooms with spines are a feature of the *tooth fungi*. Mushrooms with a stalk, cap, and gills are called *agarics*; and mushrooms with a stalk, cap, and pores are called *boletes*.

This brings us to a discussion of the **stalk** (also called stem or stipe). The stalk is the structure that supports the cap. In terms of trees, the stalk would be the trunk that holds the canopy. Stalks don't always have to be in the center of the cap; they may be off-centered (or eccentric). They can be long or very short. The bottom of the stalk, at the ground level or below, can be round like a bulb or straight. When cut open, the inside (or context) can be firm, hollow, or spongy. Texture is important in identification. Remains of other parts of the mushroom, such as the volva or partial veil, can be seen on the stalk. The few mushrooms that do not have stalks are called *sessile*.

The **veil** is a thin tissue that covers some part of the immature mushroom. As it begins to mature, the veil breaks apart and pieces of it can be seen on the cap and the stalk. A **universal veil** covers the entire mushroom, whereas a **partial veil** covers only the gills of the mushroom. On a mature mushroom, remnants of the universal veil may remain on the top of the cap as warts or as a single patch; at the base of the stalk, there may be scales or rings of tissue, or a large cup or boot of tissue called a **volva**. On a mature mushroom, remnants of the partial veil may remain as an **annulus**, or ring on the stalk.

The whole point of this structure is for reproduction, and the end result is **spores**, the tiny dust-like reproductive propagules of the fungus. They are borne on the walls of the gills, teeth, or pores of the mushroom. As mentioned above, the spore color may vary from one group of mushrooms to another, and it may be helpful in identifying your mushroom. Some colors are common, including white—many groups of agarics have white spores, and many boletes have brown spores. In contrast, only a few groups of mushrooms have pink, and only one species has green. To check the color of a mushroom's spores, you can make a spore print. Simply place a mushroom cap on (or suspend it just above) paper, cover the mushroom with a bowl (creating a chamber that will keep the mushroom humid), and leave it for a few hours or overnight.

Cortinarius glaucopus spore printing on paper

Disclaimer: This is not an all-inclusive resource for determining differences between mushrooms. Other references and experts should be sought out to help determine specific types of mushrooms. No mushroom should be eaten unless edibility is absolutely certain. Assume that all mushrooms are poisonous until proper identification is made. Even at that point, eat at your own risk!

Mushrooms: Identification

If you learn and follow the basic steps in this book, you will be well on your way to proficiency with wild mushroom identification. Although this book is suitable for beginners, even experts follow these same steps when trying to figure out a wild mushroom. Before you pick that mushroom: Be observant. All factors are clues that you can use to make an identification. What is the habitat? Temperate forest, prairie, desert, rainforest, Mediterranean, etc.? Are there trees nearby, and are they conifer, broadleaf, or a mix?

What is the substrate, meaning, is the mushroom growing from duff on the forest floor, fallen rotting wood, standing live wood, a lawn, or flowerbed mulch in an urban setting? Maybe take a few photos and jot down a few notes.

What is the morphology of the mushroom? That is, what are its features? Does it have a stem and cap with gills, or a stem and cap with pores? Is it round with none of those features? Does it have a shelf or shelves growing from the side of a tree, a simple club shape, or the appearance of undersea coral?

If it has a stem, make every attempt to examine the base of the stem—it may be buried—and note any features: Is it enlarged at the base or does it sit within a cup or boot-like material?

Finally, one of the most informative features of mushrooms (if present) are the spores. Spores are the tiny, dust-like reproductive propagules of mushrooms (similar to the pollen of plants). An individual spore is far too small to see with the naked eye. But for the purposes of mushroom identification, we mostly need only to see what color they are. When a mushroom deposits lots of spores onto a surface, such as the soil beneath a mushroom cap or other overlapping mushroom caps, you will see a spore print. Sometimes you can tell what color the spores are simply by looking at the gills or pores on a mature mushroom specimen. Spore colors can range from light colors such as white, cream, or yellowish, to light pink or salmon-pink, to darker colors such as brown, black, or purple-black. One important species worldwide has green spores.

A final note on mushroom identification: This book is a starting point and does not come close to showing all the mushrooms in your area.

Aleuria aurantia

Most of the major groups of mushrooms in your area will be featured, however. This means that you will likely be able to determine most common species in your area, or very close relatives.

No book has every species of mushroom in a given region. Furthermore, many mushrooms have not even been described and still await a formal name. And experts get stumped too.

Part 2.1. Agarics with Pale Spores

Spore color can vary greatly among the gilled mushrooms. The most common colors are pale. The colors range from absolute white, yellow, or pale orange, to various shades of ochre, to cream or buff.

Part 2.1.1. Parasols and Shaggy Parasols

Parasol and shaggy parasol mushrooms are mostly found growing out in the open, often in parks or near dwellings. They favor mulch and leaf litter, as well as grassy areas. All are saprobes and rot grassy or woody debris. This group includes some of our largest stalked mushrooms, exceeding 12 inches (30 cm) in height and with caps up to 10 inches (25 cm) in diameter.

All members feature a stalk with ring and gills that remain white at maturity, with one very notable exception: *Chlorophyllum molybdites*, the green-gilled lepiota, which is quite toxic and is one of the most common causes of mushroom poisonings worldwide. It is our only mushroom with green spores. This group contains edible species as well, but they could be confused with small woodland lepiotas (see below), all of which are inedible or toxic, and amanitas, some of which are deadly. You may hear the parasols and shaggy parasols referred colloquially as "lepiotas" because they were once considered species of *Lepiota*.

Parasol mushrooms of eastern North America and Europe include *Macrolepiota procera*, which has a very long stalk that is scaly brown with a ring that is visibly loose (you can slide it up and down), and *Leucoagaricus naucinus* (also called *Leucoagaricus leucothites*). The latter species is known as the smooth lepiota, for it lacks scales altogether and discolors gray or brownish where handled. This mushroom is typically found in grassy areas (often without trees nearby) and is similar looking to deadly white amanitas and field mushrooms (*Agaricus* species). Deadly white amanitas (destroying angels) also have white gills but always occur with trees; field mushrooms have gills that become a chocolate-brown color upon maturity.

The shaggy parasols of western North America bear a close resemblance to parasols but feature smooth white stalks (with a ring) and scalier or shaggy caps. The flesh of all shaggy parasols will discolor (reddish, orange, or brown) upon handling or slicing. The base of the stem of all mushrooms in this group will be bulbous, but unlike some amanitas, will never have a cup (volva); look for fuzzy white mycelium and debris attached to the stem base. *Chlorophyllum brunneum* and *C. olivieri* are similar (the latter is a bit more greenish in color and is more typically found in woodland) and are common in western North America. *Chlorophyllum rachodes* is common in Europe and the East, but much less so in the West. The green-gilled lepiota (*Chlorophyllum molybdites*) is a large white mushroom that often appears in large fairy rings on lawns in warm and especially hot climates. It bears a close resemblance to other species of this group. And note that the absence of green spores is not foolproof: The gills will turn greenish-gray only once the specimens are very mature or over the hill.

Part 2.1.2. Woodland and Small Lepiotas

There are many small lepiotas that resemble parasols and shaggy parasols. But members of this group are typically less conspicuous; they mostly reside in forests and are smaller in stature: 2–4 inches (5–10 cm) tall with caps

Macrolepiota procera

Leucoagaricus naucinus

Chlorophyllum brunneum, left, and *C. rhacodes*, right

Chlorophyllum rachodes

Chlorophyllum molybdites closeup of green gills

Chlorophyllum molybdites fairy ring

Lepiota magnispora

Lepiota lilacina

Leucocoprinus birnbaumii

Lepiota cristata

Leucocoprinus birnbaumii close-up

about 2 inches (5 cm) in diameter. A few may be quite showy, with shaggy caps and stems. Caps for this group are typically darker colored at the center. None should be considered edible; some contain the same deadly amatoxins that are found in some amanitas. All are saprobic in plant debris; the smooth lepiota occurs on lawns. All have white gills, a stalk with ring, and a somewhat bulbous stalk base resembling amanitas.

Lepiota magnispora and *L. clypeolaria* are quite showy and have shaggy stalks; they are found across all North America and Europe and are all but indistinguishable. *Lepiota cristata* and *L. subincarnata* have a similarly wide range but feature smooth stems with similar colors as the reddish cap disc. *Lepiota lilacina* is smaller and has lilac tones, and it seems to favor mulch in urban settings.

One other member of this group is worth mentioning. The flower pot lepiota (*Leucocoprinus birnbaumii*) is a small lemon-yellow mushroom found in mulch, flower beds, and grass. The mycelium is often present in nonsterile potting media, and the mushrooms subsequently develop in flower pots and greenhouses. The upper part of the stem is thin and fragile, and the base is swollen. Although it is one of the most attractive mushrooms, this species is poisonous. Older books may list it as *Lepiota lutea.*

Part 2.1.3. Amanitas

Perhaps the most famous of all groups of mushrooms is the genus *Amanita.* These mushrooms are some of the most recognizable from their appearances in video games, movies, and folklore. *Amanita* is an enormous genus and is common in every type of forest around the globe. Nearly every species of *Amanita* is mycorrhizal, and they can occur with a broad range of conifers and hardwoods.

The genus is divided into many different sections. Each section has its own set of defining physical characteristics. The shape and structure of the base of the stem is one of the most important features for identifying amanitas. An immature, developing amanita is enclosed in a membrane called the universal veil. As the young mushroom, or "button," draws in water, it starts to expand. This expansion causes the universal veil to rupture. In doing so, it can stretch and tear in many different ways. The mushroom will then grow to full size, and the remnants from its original wrapper can help us to determine which section of the *Amanita* genus we are looking at. This can range from a well-defined cup to a very fragile or ephemeral sleeve. In many cases, the slow explosion of a rupturing universal veil will leave tissue on the cap that is referred to as "warts."

Many mushrooms in this genus also have a second veil on the underside of the cap, protecting the developing gills. It surrounds the stem and is attached to the edge of the cap. This membrane is called the partial veil. As the cap expands, this membrane breaks, revealing the mature gills. When the partial veil breaks, it detaches from the edge of the cap, leaving a ring or a skirt on the upper stem of the mushroom.

In general, Amanitas are pretty large mushrooms with caps about 4–6 inches (10–15 cm) wide and stalks 6–10 inches (15–25 cm) long for many species.

Though many amanitas are edible, there are toxic species and a handful that can be lethal. The most familiar deadly species are the yellow-green death cap (*Amanita phalloides*, see page 18) and the all-white destroying angels (a group of similar species including *Amanita bisporigera,* see page 18). These mushrooms produce amatoxins, which block DNA transcription, and thus cell division; death is usually due to organ failure, most often the liver and

Amanita jacksonii

Amanita phalloides

Amanita bisporigera, destroying angel

Amanita muscaria var. *flavivolvata*, West

Amanita muscaria var. *formosa* (*guessowii*), East

Amanita vaginata

Europe are both red; *A. calyptrata* of the West is more brownish) are popular edibles for the very experienced mushroom collector. The grisette, *Amanita vaginata*, is gray with a smaller cup, and notable for lacking a ring on the stem. Hundreds of species of *Amanita* exist, and all are worth a photo when encountered.

Part 2.1.4. Chanterelles

Chanterelles are among the most prized of edible mushrooms. Most species are a yellowish color, but they may also be white, pinkish, purplish, reddish, brownish, or black. All are vase-shaped and mycorrhizal, most often with oak species. Chanterelles range from about 1 to 4 inches (2.5 to 10 cm) in size, but some species on the West Coast can be much larger.

There are two genera in this group; species of genus *Cantharellus* are solid, including the stem; species of genus *Craterellus* are hollow or tube-like, including the stem. The "gills" (not actually true gills) are usually no more than small ridges. Pretty much all yellow chanterelles at one time were called *Cantharellus cibarius*, the "true" chanterelle now thought to be restricted to Europe. (The one pictured is from Italy.)

kidneys. This group of amanitas cause more mushroom poison–related deaths than any other mushrooms. The death cap is annually responsible for at least 90% of all mushroom poisoning deaths worldwide.

Other very recognizable mushrooms from this genus include *Amanita muscaria* and all of its varieties, red-capped in the West and Europe, yellow-capped in the East. All varieties have white warts on the cap; remnants of the universal veil. Known as the fly agarics, these mushrooms produce different toxins (primarily ibotenic acid) that can cause serious gastric upset. The brightly colored Caesar amanitas (*Amanita jacksonii* of the East and *A. caesarea* of

Cantharellus cibarius, Italy

Cantharellus roseocanus

Cantharellus cinnabarinus

Hygrophoropsis aurantiaca is on the right

Craterellus ignicolor

Craterellus cornucopioides

The most common yellow chanterelle of eastern North America is *Cantharellus flavus*, from western North America is *Ca. californicus* and *Ca. formosus*, from the South is *Ca. tenuithrix*. All are fairly large, gregarious species that occur with oaks. Another yellow species, recognizable by its peach-colored tones, is the rainbow chanterelle, *Cantharellus roseocanus*; it ranges from the West into the Midwest. The smooth chanterelle (*Cantharellus lateritius*) of the East has a nearly smooth, spore-bearing surface.

Chanterelles could be confused with the poisonous jack-o'-lantern mushroom, *Omphalotus* species, which has a similar funnel shape, color (yellow-orange), habit (it grows under broadleaf trees, especially oak), and season (late summer). The false chanterelle, *Hygrophoropsis aurantiaca*, is very similar in appearance, though it has true gills that fork repeatedly (see the photo comparing the two; the chanterelle is on the left).

There are very small chanterelles, many of them poorly known. The cinnabar-red chanterelle, *Cantharellus cinnabarinus*, is strikingly pretty to see and fruits in profusion, but it is not worth mentioning as an edible.

Craterellus cornucopioides, the black trumpet, is known across North America and Europe and is highly prized. When dried, it smells of truffles or Romano cheese, hence its French name, which translates to "poor man's truffles."

Craterellus fallax is similar to *Cr. cornucopioides* and can be distinguished only by spore print; *Cr. fallax* has salmon-buff spores while *Cr. cornucopioides* spores are white. *Urnula* species look similar to black trumpets but are found in early spring.

There are also small species of *Craterellus*, some of them notable, such as the yellow-foot chanterelles group, including *Craterellus tubaeformis*, with a typically gray hymenium, and *Cr. ignicolor*, which is entirely yellow. Both are decent edibles and found across North America; *Craterellus lutescens* of Europe is similar.

Part 2.1.5. Collybioid Mushrooms

The term *collybioid* refers to mushrooms that all share a common set of physical features. They were all formerly placed in the genus *Collybia* but have since been separated into several genera. *Collybia* now only has a handful of species. We still use the term to describe certain attributes.

Collybioid mushrooms are small, fragile mushrooms with white spore prints. They occupy wooded areas, decomposing wood, and other forest debris. The largest and most fre-

quently observed genus is *Gymnopus*. *Gymnopus dryophilus*, one of the most common mushrooms of North America and Europe, can be seen almost anywhere and at any time during the year when it's warm enough. The cap is about an inch (2.5 cm) wide, is reddish-brown, and it often has zones of different color that change as the mushroom dries. The stem is straight and slightly grooved, and gradually tapers to an enlarged base. The gills are white to pinkish with a white spore print. Note that *dryophilus* means "oak-loving," but it is just as likely that you will see this common collybioid in any hardwood or conifer woodland setting, even where no oaks are present. This mushroom is edible, but don't pick it from polluted areas.

Two more common collybioids are the small parasitic/saprobic *Collybia cookei* and *C. tuberosa*. Both are very small and white, and both grow in groups on decaying matter, including decaying mushrooms (and might also be directly parasitic on the mycelium of living mushrooms, especially mushrooms in the genus *Russula*). The fruit bodies arise from small underground tubers, which can be found by digging in the duff beneath the mushrooms.

Gymnopus dryophilus

Collybia tuberosa

Collybia cookei is noted by yellow sclerotia and *Collybia tuberosa* by reddish-brown sclerotia, reminiscent of apple seeds. Both species are widespread across North America and Europe.

There are also collybioids that occur on wood. *Flammulina velutipes* is also known as velvet foot and the winter mushroom; it is one of the few mushrooms that can be found during the winter months in the northern hemisphere. This mushroom is usually found in groups on dead wood. In northern North America and Europe, it is frequently seen in the spring on dead elm (*Ulmus* species). The upper surface of the cap is orange-brown and tacky to slimy. The name velvet foot comes from its velvety stalk, which is dark brown or black, particularly in the lower portion. The gills are adnexed or notched and unless examined closely, may appear to be free. This species is widely cultivated in Japan, where it is known as enokitake. The cultivated type looks nothing like its wild counterpart, more closely resembling vegetable sprouts.

Caution: Without careful examination, *Galerina marginata* could be mistaken by beginners for the velvet foot. Both appear throughout the year (especially late fall), are of similar size, occur on the same substrate (dead, deciduous wood), and have brownish, tacky to viscid caps. *Galerina marginata* is potentially lethal if eaten, and beginners should become familiar with both species to avoid making a deadly mistake.

In late spring, after the morel season, there is a brief period when the mushrooms seem to disappear from the forest—all except for *Megacollybia rodmani* (*M. platyphylla*). It is easy to identify and edible, but it is not considered very good table fare, as it has an odd, somewhat unpleasant taste. Called the platterful mushroom, it has a large (4–6 inch [10–15 cm]), grayish to brown cap and a fibrous stem. It is most commonly found growing from stumps or large pieces of decaying wood in eastern North America and Europe. Upon careful inspection, you will note very tough hyphal cords at the base of the stalk, and these may extend several feet to its rotting wood substrate (a log or roots), giving it the appearance of being terrestrial. It resembles some of the larger species of *Pluteus*, but they have free gills and pink spores. *Megacollybia* has a white spore print.

One group of mushrooms with a unique structure is the rooting collybioids. These mushrooms all feature a long "taproot" that can be most of the mushroom's overall length. Though people refer to this as a root, it is not a root at all. The mushroom does not start growing downward from ground level. Instead, it starts deep below the surface of the ground and expands upward, which gives it the appearance of having a root. Rooting collybioids have gone through many taxonomic changes over the past few years and all currently reside in the genus *Oudemansiella*. The caps of some of the larger species,

Oudemansiella radicata in situ

Flammulina velutipes

Oudemansiella radicata roots

Flammulina velutipes, cultivated

Megacollybia rodmani

Megacollybia rodmani with long hyphal cords

such as *Oudemansiella radicata* (*Xerula radicata*) of the East and Europe (3–5 inches [8–13 cm]), are sometimes collected for the table, but they have an odd, fishy flavor that is not generally viewed as favorable. The cap can be easily recognized by its radially arranged wrinkles, and the gills are typically wide and distantly spaced. The top half of the stem is often darker, featuring a pattern that resembles snakeskin. Once an *Oudemansiella* is found, pull it very gently to see how much "root" you can retrieve.

Part 2.1.6. *Clitocybe* Group

A frequent encounter in the forest will be various members of the genus *Clitocybe*. All *Clitocybe* species are saprobic and are commonly found on decaying wood or forest litter. They may be found in hardwood or conifer forests, often in great numbers. The genus is a difficult one to break down accurately. It is large, and many members are rather plain-looking and medium-sized, with 2–4 inch (5–10 cm) caps. They lack unique features to separate them from each other. The typical spore color will range from perfect white to yellowish, although some do have a pinkish hue.

Most members of the genus have distinct odors. They are difficult to describe, but they all have a similar sour element. Once you recognize this element, you can identify mushrooms as *Clitocybe* just based on their odor. One species that has an aroma with a little extra is *Clitocybe odora*. Along with the sour notes of the typical *Clitocybe* fragrance, it has a scent strongly resembling anise. Some have a funnel shape and club-shaped stem (that feels spongy), such *Clitocybe* (*Ampulloclitocybe*) *clavipes*. Both species are widespread across North America and Europe.

The jack-o'-lantern is a bright orange to brown-orange mushroom found growing from stumps or buried wood. It grows in clusters, and all the stems are fused at the base. These mushrooms can be very large, with caps to 8 inches (20 cm) wide and stalks 12 inches (30 cm) long. The jack-o'-lantern is named for its pumpkin color and the fact that it is found in October, but also because the gills and other parts glow in the dark. They have a superficial resemblance to chanterelles, and beginners do occasionally mix them up.

Omphalotus illudens is the bright orange species from eastern North America and Europe (it is sometimes erroneously called *O. olearius*, but that species is European and mostly found on the wood of olive trees, as the name implies). *Omphalotus olivascens* is the species known on the West Coast of North America; it features a darker orange color with olive green tints. All species of *Omphalotus* are poisonous and produce a toxin called illudin, which causes severe gastric upset.

Beginners should become familiar with this group to prevent any confusion with chanterelles—distinguishing the two is relatively easy. The gills of *Omphalotus* are orange, relatively sharp-edged (as opposed to the blunt ridges of a chanterelle), and decurrent. They will fan like a deck of cards if you run your finger across them. Chanterelles will not. The interior flesh of the jack-o'-lantern is orange, whereas the flesh of the chanterelle is white. *Omphalotus* species grow from rotting wood, but they may be buried.

Clitocybe odora

Clitocybe clavipes

Omphalotus olivascens

Omphalotus illudens gills

Omphalotus illudens

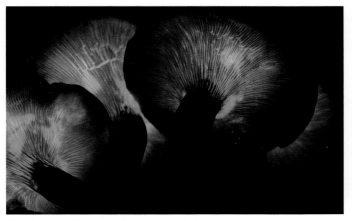

Omphalotus illudens glowing

Omphalotus olivascens

Part 2.1.7. Honey Mushrooms

Older books refer to honey mushrooms as *Armillaria mellea*. They also go by many additional common names, including shoestring mushrooms. The honey mushroom is highly variable in size (usually 2–4 inches [5–10 cm] for caps), shape, and color. Nowadays we know this variability in appearance is because honey mushrooms are actually a group of species of *Armillaria*.

One of the key identifying features of honey mushrooms is the presence of tufts of dark hairs near the center of the cap; another is the fibrous nature of the stem. *Armillaria* species cause a root and butt rot of living trees, both hardwoods and conifers. The abundance of this fungus is apparent by the widespread presence of black rhizomorphs ("shoestrings") growing on dead and dying trees and fallen logs. Rhizomorphs are darkly colored, hardened strands of hyphae, and they can grow for many feet (or miles) along the forest floor.

Armillaria mellea (*mellea* meaning "honey-colored") has a partial veil that leaves a ring on the stem. *Armillaria tabescens*, and the almost indistinguishable *A. gallica*, have ephemeral,

Armillaria mellea group

Armillaria mellea **close-up**

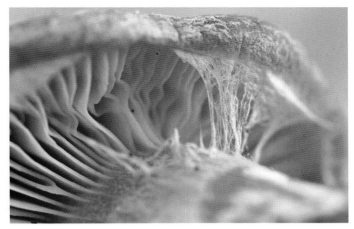

Armillaria gallica **cobweb veil**

Armillaria tabescens

cobwebby veils that leave no ring on the stem; they are called ringless honey mushrooms. The color is usually orange-brown, and, like its ringed relative, a ringless honey mushroom has tufts of dark hairs near the center of the cap, a fibrous stem, and rhizomorphs. This mushroom grows on wood, especially on oak and maple.

All honey mushrooms are popular edibles and are among the most collected mushrooms in the northern hemisphere. Their flavor can be very strong to bitter, and the stems are not really used as food, as they are very tough and fibrous.

Part 2.1.8. Matsutake Group

After most other mushrooms have disappeared for the year, the knights, as they are known, make their entrance. These are the mushrooms in the genus *Tricholoma*. They are medium to large mushrooms that are mycorrhizal with a wide range of hosts. Most of them are dull colored and less than exciting to look at.

There are very few tricholomas that are targeted for the table, but one exception is the matsutake, or pine mushroom. The name matsutake is applied to at least three species of *Tricholoma*. They are famously fragrant and coveted by chefs around the world.

Tricholoma murrillianum

Tricholoma caligatum

Tricholoma focale

Tricholoma focale, *T. caligatum*, **and** *T. murrillianum*

The major species of matsutake in North America are *Tricholoma murrillianum* in the West, *T. magnivelare* in the East (and Europe), and *T. mesoamericum* in the Southwest and Mexico. All three are pretty much indistinguishable conifer associates (pines and Douglas-fir) that prefer poor soils.

Matsutake has a very spicy odor that can be difficult to describe. It has notes of cinnamon candy and cedar and is strong enough to tingle the nose. This odor holds up with gentle cooking, leaving a unique, piquant flavor. They can be very large mushrooms, sometimes exceeding 10 inches (25 cm) in diameter.

Occasionally people will confuse close relatives of the matsutake for the real thing. These "notsotakes" may closely resemble matsutakes in appearance, but not flavor. *Tricholoma caligatum* is one of these species. This beautiful mushroom has a dry, white cap with brown or reddish-brown fibers and scales. The annulus has a jagged margin and is often turned upward. *Tricholoma caligatum* is commonly found with oak trees and, depending on the region, can have a matsutake-like odor. More often, though, it smells really foul (like urine). It is not poisonous, but it is typically bitter and therefore not palatable.

Tricholoma focale (older books may use *T. zelleri*), is every bit as beautiful but with orange and olive tones, and it is also not collected for the table. A large shaggy white *Amanita* species (Smith's amanita, *A. smithiana*), which is toxic, is occasionally confused with matsutake in the Pacific Northwest; there have been associated poisonings and at least one death.

Part 2.1.9. Mycenas

There are more than 250 species of *Mycena* in the forests of the northern hemisphere. They are saprobic and mostly found growing on decaying wood, but some are found on other forest litter. They are delicate, bell-shaped mushrooms, rarely exceeding an inch (2.5 cm) in diameter. Mycenas are often overlooked because they are too frail to be considered a target for the table. They commonly grow in large troops, sometimes with hundreds of individuals.

There are many species of *Mycena* that are tan to brown or gray-brown and are commonly seen across North America and Europe, including *Mycena inclinata*. These common saprobes feature a nearly conical cap that eventually expands to nearly flat. The cap features radial striations, becoming more distinct towards the edge of the cap. A close examination of the area between the gills reveals prominent crossveins. The base of the stem is usually covered in white mycelium, giving it a fuzzy look. This mushroom has the general appearance of an inky cap, but the gills are white to grayish pink (never dark), and spore print is white.

Not all mycenas are white or drab shades of brown. A definite eye-catcher of the genus is *Mycena leaiana* (see image page 30). It is one of the most colorful—a bright orange mushroom with dark orange to red gill edges. Although the individual mushrooms are small, they are easy to find because they often grow in groups of five or more. *Mycena leaiana* is common in the spring when few other mushrooms are out. Take only photos of this one, as it is poisonous.

Another member of the spring cast is *Mycena haematopus* (see image page 30). This character is commonly known as the bleeding fairy helmet. The small, pale to reddish-brown mushroom is easy to identify by the dark red juice it exudes when it is cut or broken; the caps show reddish bruising where damaged. It is similar in size to *Mycena leaiana*. Both spring mycenas are common on hardwood stumps.

Mycena inclinata

Mycena crossveins

Mycena haematopus

Mycena epipterygia

Mycena leaiana

A little bit later in the year, we start to see *Mycena pura* and *M. rosea*. These small mushrooms are similar in size and shape to other *Mycena* species; however, these pinkish mushrooms grow on the ground rather than on wood. The color ranges from pink to lilac to purple, but fades to brownish-gray with age.

Some of the tiniest mycenas are the most photogenic. They include the hot pink *Mycena* (*Atheniella*) *adonis*, the walnut-loving *M. luteopallens*, and *M. epipterygia*, whose bright yellow stem almost glows. Like others in this genus, you will find crossveins between the gills of all these mushrooms.

Part 2.1.10. Fairy Ring Mushrooms

A mushroom in the genus *Marasmius* can be easily mistaken for a *Mycena* at first glance. Once we handle the mushroom, however, we will see that *Marasmius* mushrooms are not nearly as delicate. Some of them can even be described as wiry. They are most common in midsummer and can thrive in dry conditions.

A common representative of the genus is *Marasmius capillaris*. It frequents the leaf litter under oak trees and can appear in troops numbering in the hundreds. They are tiny, typically less than a half inch (1 cm) in diameter.

Large for its genus is *Marasmius oreades*. Commonly called the Scotch bonnet or fairy ring mushroom, this saprobe can be found in lawns, pastures, or other grassy areas across North America and Europe. It has a very rubbery stem, but the cap is tender and edible. During dry conditions, this mushroom can completely dry out, then rejuvenate and magically spring back to full size with the return of moisture. This ability, combined with the mushroom's curious habit of growing in rings, led many in the days of old (and perhaps even today) to believe that it was the result of fairies. It can resemble other mushrooms and could be confused with toxic members of *Inocybe* or *Clitocybe*. Becoming familiar with similar mushrooms is recommended before this one is collected to eat.

Marasmius oreades fairy ring

Part 2.1.11. Laccarias

Nearly every forest system in the northern hemisphere will have populations of *Laccaria* during the late summer and autumn. They vary slightly in size (usually 1–4 inch [2.5–10 cm] caps) but they all have waxy, pinkish-brown to purple gills and a white spore print. None of the laccarias are known to be poisonous, but many people avoid them because some are difficult to identify to species.

Many species of *Laccaria* are overlooked because they are small and barely protrude above the forest litter, but they are all worth a look. The very small *Laccaria amethystina* is brilliantly colored and found across North America and Europe. All parts are a rich purple color that is sure to fascinate.

Some of the larger *Laccaria* are sought out as edibles. The most common is *Laccaria ochropurpurea* of the East and Europe, and *L. amethysteo-occidentalis* of the West. They are the largest of the genus. They are fairly easy to identify by their dry, purple-gray cap and thick, purple gills, but they are often confused with potentially toxic *Cortinarius* species. It is easy to distinguish laccarias from *Cortinarius* based on its white spore print. (*Cortinarius* species have rust-brown spores.)

Laccaria amethystina

Laccaria ochropurpurea with *Cortinarius* and arrows showing similarities

Part 2.1.12. Milk Mushrooms

Lactarius and *Lactifluus* are the two genera that make up the milk mushroom group. They are closely related to *Russula*, exhibit many of the same traits, and are medium-sized (about 2–4 inches [5–10 cm] for caps). Although they are fragile, terrestrial, and have a similar stature to *Russula,* the milk mushrooms exude latex when they are cut or damaged. This latex is often the consistency of milk, but the color can vary widely. From the neon yellow of the poisonous *Lactarius vinaceorufescens* to the electric blue of the edible and cosmopolitan *Lactarius indigo*, the broad range of colors can be eye-catching. The color of the latex is a very important feature in identifying *Lactarius* and *Lactifluus*. The latex of some species changes color as it dries, and it is important to take note of this color change.

All milk mushrooms are mycorrhizal. They can be found with both conifers and hardwoods and are common mushrooms throughout summer and fall.

Most "milkies," as this group is called, are edible, but only a few are worthy of collecting. The orange milk members of the *Lactarius deliciosus* group are without doubt most famous; they are known across North America and Europe. The milk of *Lactarius deliciosus* remains orange, while that of *L. deterrimus* changes from orange to green.

Lactifluus volemus, often called the bradley, has a unique flavor and is frequently harvested for the table. It has an odor similar to that of shrimp or crab, with an additional sweet element. The latex emerges bright white, but quickly changes to brown. It can actually leave stains on your skin or clothing.

The candy caps, a group that includes *Lactarius camphoratus*, *L. rubidus*, *L. rufulus*, *L. fragilis*, and similar looking species, are small (1–2 inches [2.5–5 cm]), reddish mushrooms that exude a clear watery milk. They are not

Lactarius indigo

Lactifluus volemus

Lactarius deterrimus

Lactarius camphoratus

Lactarius glyciosmus

Lactarius deceptivus

actually sweet tasting, but they have a smell of fenugreek, maple syrup, or burnt sugar. Although the smell is faint when the mushrooms are fresh, the aroma gets stronger when they are dried. They are one of the few mushrooms that work in dessert recipes. They are found across North America with conifers or hardwoods, depending on species.

Many milk mushrooms, including *Lactarius deceptivus*, can have a sharp to very acrid hot taste and are totally unpalatable. *Lactarius deceptivus* always has a hole in the cap and white milk. Even the drab-looking little *Lactarius glyciosmus*, common across North America and Europe, has a surprise—it has a wonderful scent of coconut, but is not considered edible.

Part 2.1.13. The Brittle Gills

Russula is a large genus of medium-sized (2–4 inches [5–10 cm] wide) terrestrial mushrooms. They are on every forest floor across the globe, but don't get the attention that other groups of mushrooms get. Their bright colors catch the eye of the beginner, but the fascination fades as their limited value is discovered. They are not often sought out as table fare and do not have large, showy individuals to draw in photographers. They are famously fragile, and some species can barely be handled without breaking.

There are some palatable russulas, but there are also rules to follow before we decide to add them to our dinner plate. Rule number one is, do not eat any *Russula* that has a spicy or acrid taste. Take a nibble of the edge of the mushroom and crush it between your teeth. Use your tongue to push the sample out of your mouth. This is a perfectly safe practice with all mushrooms, as a poisonous mushroom (even deadly species) has to be swallowed to do any damage. Wait 30 seconds. Some reactions are instant, while some are quite delayed. If you detect the tingling of hot peppers or acid taste, discard the mushroom. The second rule is, do not eat any *Russula* that discolors when sliced. This can mean gray to black, or pink to red to black. The third rule is, avoid any *Russula* that smells like almonds, maraschino cherries, or bleach. These have been known to cause some gastric discomfort.

None of the russulas in Europe or North America are deadly poisonous, but there are some suspects in this group that inhabit the forests of southern Asia. And there are a couple of notorious *Russula* species in both Europe and North America that gave the genus a bad name in the past. The European mushroom is *Russula emetica*. As the name suggests, this mushroom

Russula peckii

may cause vomiting if eaten, especially if eaten raw. The North American version is *Russula silvicola*; it has a bright red cap, white gills, and white stem. They both have an incredibly hot taste which makes them easy to avoid if you do the taste-and-spit test. The similar *Russula peckii* has blush on the stalk.

Russula cyanoxantha

Some of the better edibles of this group include the greenish *Russula cyanoxantha* and *R. virescens*, and the purple-red *R. xerampelina*. These are worth learning, as they are sometimes abundant and easy to feel comfortable with. *Russula xerampelina* is known as the shrimp russula for its flavor.

Part 2.1.14. Lobster Mushroom

The lobster mushroom is actually two species of fungi wrapped up in one. Various species of *Russula* and *Lactifluus* host the crippling lobster fungus, *Hypomyces lactifluorum*. When the mycelium of these host mushrooms become infected with this parasite, the fruiting bodies emerge as disfigured clumps barely resembling the typical mushroom of the host. They are called lobster mushrooms because of their orange color, which resembles the color of a boiled lobster.

None of the host mushroom species are typically considered edible (some are acrid), but the lobster fungus changes this. It mellows this spicy, unpleasant taste and leaves behind a mild, very firm edible mushroom. The texture of the lobster mushroom is second to none, remaining firm after cooking. Lucky are those who find massive "lobsterized" *Russula brevipes*—these are the largest of brittle gills, and they are widespread across North America and Europe.

Russula brevipes and lobster mushroom

Hypomyces lactifluorum

Part 2.1.15. Wax Caps

Wax cap is a general name used for forest and grassland mushrooms that have a waxy feel. The genera *Hygrocybe*, *Cuphophyllus*, *Humidicutis*, and *Hygrophorus* are the mushrooms most referred to as wax caps or waxy caps (although older books may use only the genus *Hygrocybe*).

Wax caps can be very striking, especially the genus *Hygrocybe*. These small, glossy mushrooms can be bright yellow to deep crimson and every color of orange in between. They often grow in large troops, coloring the forest floor just before the leaves fall in autumn. None of the wax caps are large mushrooms (they get up to about an inch [2.5 cm] or so), but they still catch the attention of forest visitors.

Hygrocybe punicea is one of the largest of the colorful wax caps. It is reported to be edible, but it is not often collected because of its tendency to disappear when cooked. *Hygrocybe conica*, the witch's hat, is similar but turns black upon handling. Both are common across North America and Europe. And one of our few green mushrooms is the parrot mushroom, *Gliophorus psittacinus*.

A wax cap with a completely different guise is *Hygrophorus russula*. This mushroom gets its name from the fact that it resembles something from the genus *Russula* more than anything it is actually linked to genetically. It is also much larger than other wax caps, occasionally reaching 8 inches (20 cm) in diameter. It is edible and opinions on its quality differ. It has an excellent texture and the flavor can vary from mildly sweet to earthy. It appears in the summer and fall and can be found in great numbers.

Hygrocybe punicea

Hygrocybe conica

Hygrocybe (Gliophorus) psittacinus

Hygrophorus russula

Part 2.1.16. *Xeromphalina* Species

If you discover hundreds of small, orange mushrooms covering a rotten stump, they are likely to be one of a handful of species of *Xeromphalina*. The most commonly encountered is *Xeromphalina campanella* of North America. It can be found in enormous numbers on the decaying stumps of various conifers. Close inspection will reveal tufts of orange fuzz at the base of the stem, yellow gills that are decurrent, and crossveins between the gills. The caps of all *Xeromphalina* are some shade of orange, but the stems are typically darker, sometimes even black.

Xeromphalina kauffmanii is nearly identical to *Xeromphalina campanella*, but it is found on decaying hardwood stumps. All species are saprobic and can be found in all forest types, but they do seem to be less common in Europe. Most species of *Xeromphalina* are less than an inch (2.5 cm) in diameter, though a couple of species may reach 2 inches (5 cm) across.

Xeromphalina campanella

Xeromphalina campanella

Part 2.2.
Agarics with Pink Spores

Pink is a common spore color for agarics, but the word *pink* is used loosely. The color of the spores can be pinkish tan, salmon-pink, or a dull orange.

Part 2.2.1. The Shield Caps

Pluteus is a fairly large genus of wood-rotting mushrooms, common to all types of forests. They are most easily identified by their distinctly free gills, which turn pinkish tan at maturity. Most of the mushrooms in this genus are small and delicate. There are a few species that are larger (4–5 inches [10–13 cm] for caps) and a bit sturdier, the most common being *Pluteus cervinus*. It is known across North America and Europe. Also called the deer mushroom, *Pluteus cervinus* is harvested by some for the table, but it has an odd, radishy taste and is not viewed as a high quality edible.

There are no known poisonous species of *Pluteus*, but most are fragile, watery, and not worth collecting. In the early spring when nothing else seems to be active, members of the group can still be found. They are quite tolerant of cool temperatures and dry conditions.

Volvariella and *Volvopluteus* are less well known but closely related to *Pluteus*. An uncommon, but always exciting find is *Volvariella bombycina*, easily identified by its fuzzy, bell-shaped cap and its large, papery volva. It fruits from live trees, predominantly from injuries on maple trees (*Acer* spp.), but it can be found growing on many other hardwood hosts. It can be large in comparison to other gilled mushrooms that grow on wood, with caps reaching a diameter of 8 inches (20 cm). Closer inspection will reveal long, elegant gills that turn pink with age. The gills are distinctly free from the stem.

Pluteus cervinus showing top of cap and pink gills

Volvariella bombycina

Volvariella speciosa (*Volvopluteus gloiocephalus*)

This can be difficult to see because of the bell-shaped cap. *Volvariella bombycina* is edible, but not often eaten. It is of more interest to photographers than people seeking it out to eat.

Volvariella speciosa (*Volvopluteus gloiocephalus*) can be equally large, but it occurs from rotting debris in soil. It is also known across North America and Europe. It, too, is edible but can be foul-tasting, depending on region. **Caution:** Before the gills mature and take on the pink spore color, this one can be confused with deadly destroying angels.

Part 2.2.2. Entolomas

Entoloma is an enormous genus of small- to medium-sized saprobic mushrooms (caps 1–2 inches [2.5–5 cm]), which can be found in a variety of environments. There are more than 10 other genera that closely resemble *Entoloma*, and some guides include them as such. The genus is identified by having pink spores and attached gills. They are usually found on the forest floor, but some species may appear on wood. Many entolomas are toxic and can cause severe gastric distress. Most species in this group of mushrooms are dull-colored, including the cosmopolitan *Entoloma vernum* (*Nolanea verna*), one of the first agarics to emerge in the spring. But there are occasional exceptions for blue, yellow, and salmon-colored species.

There is one *Entoloma* that breaks a lot of rules: *Entoloma abortivum*. It is not a saprobe, but rather a parasite. It attacks the mycelium of mushrooms in the genus *Armillaria*. This causes the *Armillaria* mushrooms to abort. Instead of developing correctly, they emerge as odd-shaped, whitish clumps. These clumps are popular edibles in some regions, known by the common names lamb's ears and shrimp of the woods. These names are descriptive of their physical features and have nothing to do with the flavor of the mushroom.

Entoloma abortivum

Entoloma vernum

Clitocybe nuda

Clitopilus prunulus

Part 2.2.3. Blewits and Millers

The wood blewit, or *Clitocybe* (*Lepista*) *nuda,* is one of a few clitocybe species that produce pink spores rather than white or yellow. Some people believe these pink-spored **clitocybe**s belong in their own genus, *Lepista.* The wood blewit behaves like any other *Clitocybe.* It is a forest saprobe that can be found under conifers and hardwoods. It is often buried in the leaf litter, barely making itself visible by pushing up debris in small mounds. When one is found, start digging around for others. They are rarely alone.

In colder climates across North America and Europe, blewits are some of the last edible mushrooms to be found before winter. They are fairly large (3–5 inch [8–13 cm] caps) and they make good edibles, but identifying them can take a bit of practice. They start out purple and then mature to brown, and for this reason they are often confused with similar-looking species of *Cortinarius* (which has rust-brown spores) or occasionally some species of *Laccaria.* It is easy to separate these two genera on spore color

alone. Blewits have a faint but distinct fruity or citrusy odor. Though this mushroom is commonly collected for consumption, it is mildly toxic when raw; it should be cooked thoroughly before eating.

The miller, *Clitopilus prunulus,* is one of only a handful of mushrooms in the genus *Clitopilus.* All members of the genus produce a brownish-pink spore print. The miller gets its name from its odor of wet grain or yeast. The smell is strong and sour, but pleasant. It is a saprobic mushroom and can be found growing in the grass around hardwood forests and occasionally in mixed conifer forests. At maturity, the mushroom's gills turn from a dull gray color to pinkish brown. The gills are decurrent and the edge of the cap is typically quite wavy. This mushroom is a great edible, but there are several species of *Entoloma* and *Clitocybe* that can look similar. Mushrooms in these two genera have the potential to be poisonous, so use extra caution when identifying the miller.

Part 2.3. Agarics with Brown Spores

Many agarics have brown spores. These mushrooms all exhibit colors from dull brown to rust or reddish brown. Often these colors can be seen on the stem of the mushroom as spores accumulate.

Part 2.3.1. The Webcaps

Cortinarius could be the largest genus of gilled mushrooms on the planet. Though difficult to identify to species, identifying a "cort" to genus can be quite easy. Webcaps are medium to large mushrooms (with 4–8 inch [10–20 cm] caps) with rusty-colored spore prints, but they are most well known for their cortina. The cortina is a cobwebby veil that protects the developing gills of the mushroom. As the mushroom draws in water and the cap expands, the cortina stretches and breaks. Many of the fibers will cling to the stem of the mushroom, invisible, until they start to collect the rusty spores as they drop. This will often make the cortina visible again, along with turning the gills rusty-brown.

All species of *Cortinarius* are mycorrhizal and will always be found in the presence of trees. There are many color variations within the genus. Brown and red are common colors; one example is the cosmopolitan *Cortinarius trivialis*. There are also many lilac and purple species. *Cortinarius violaceus* is a beautiful deep purple with a dry cap, while *C. iodes* and *C. salor* are slimy and lilac-colored; all three are common across North America and Europe. *Cortinarius vanduzerensis*, from the West Coast, is recognizable by its purple stem, brown cap, and copious slime.

Cortinarius, in general, is not collected to be eaten. Though there are many edible species, there are a few poisonous ones, including a scarce few that are potentially lethal. Most of the corts that are edible are not viewed as tasty enough to be worth the trouble—or risk.

Cortinarius iodes

Cortinarius atrovirens cortina close-up

Cortinarius violaceus

Cortinarius vanduzerensis

Cortinarius trivialis

Part 2.3.2. Funeral Bells and Scaleheads

The name of this little brown mushroom says it all: The funeral bell, or deadly galerina is a small, saprobic mushroom found in nearly every forest system, on a wide variety of decaying wood. It is deadly poisonous, and beginners should become familiar with it.

Funeral bells are commonplace on wood that is already well-decayed by other organisms. They are nonselective about host wood, and they will grow on almost anything that is rotting in the forest. Funeral bells are small, fragile mushrooms (about an inch [2.5 cm] for caps) recognized by their brown cap and gills, and by their growth in groups. It is abnormal to see this mushroom growing solitarily. The developing mushrooms have a fragile, membranous veil protecting the young gills. As the mushroom grows, this membrane breaks, leaving a discolored band and occasionally a skirt on the upper stem. This area often collects the falling spores as the mushroom matures,

Galerina marginata

Pholiota squarrosa

Pholiota aurivella

Leucopholiota decorosa

Pholiota squarrosoides

Hemistropharia albocrenulata

leaving a rusty-brown collar. The cap of *Galerina marginata* is hygrophanous; that is, it changes color as it loses or gains moisture. It will appear to be a dull, straw yellow when dry, and a deep, rich brown when water levels in the cap tissues are higher.

Scaleheads also are fairly large (3–4 inch [8–10 cm]) saprobic mushrooms and they produce rusty-brown spores. *Pholiota* is a genus filled with many ornate and showy mushrooms that grow in clusters. They are predominantly wood rotters, but a small number of species can be found terrestrially. Birch (*Betula* species) trees are a common host and will often have hundreds of the clustered mushrooms bursting through the bark. But pholiotas are not limited to birch, or even hardwoods for that matter. They can be found on a wide variety of hardwoods and conifers throughout the mushroom growing season.

Pholiota squarrosa is the most well-known of the genus; the cap and stem are both covered in dry, sharp scales. This scaly look is common in many different species of *Pholiota*. Some are dry, but others have a slimy feel to the cap, including the very scaly *Pholiota squarrosoides* and the less scaly, yellowish *P. aurivella*. All are common across North America and Europe.

Also confused with *Pholiota*: *Leucopholiota decorosa* looks similar to *Pholiota squarrosa*, but is even more ornate and produces a white spore print. *Hemistropharia albocrenulata* looks like *Pholiota* and even grows commonly on birch, but it differs microscopically and by its much darker brown spore print. *Gymnopilus* species are similar, lack scales, and often are very bitter.

Part 2.3.3. Fiberheads, Hebelomas, and Agrocybes

The fiberheads are small mushrooms, usually less than 2 inches (5 cm) in diameter. Most are dull browns and grays, but there are exceptions. *Inocybe geophylla* var. *lilacina*, for example, is a beautiful pale purple color. All inocybes are mycorrhizal with a broad variety of trees and are common in all forest types. They get their name from the dense mat of radial fibers on the cap, giving them a very fibrous texture and look. When mature, the "fiberhead" becomes apparent. As the cap ages, it splits, revealing even more of the fibrous mat. Many inocybes have a foul sour or spermatic odor; a few smell fruity or subtly of green corn. A great number of *Inocybe* are poisonous and none should ever be considered for consumption.

Hebeloma is also a genus of dull-colored, mycorrhizal mushrooms. These medium-sized mushrooms (2–4 inches [5–10 cm]) are found across the northern hemisphere in both hardwood and conifer forests. They all produce a dull brown spore print, and many of them have an odor resembling radishes (it can be quite pungent). Some may have a cortina that protects the developing gills. This veil is typically ephemeral, but it can be persistent or leave a ring zone in some species. *Hebeloma crustuliniforme* (see page 48), known as the poison pie, is commonly encountered across North America and Europe.

Inocybe sororia

Hebeloma crustuliniforme

Agrocybe praecox group

Many different species in this genus look identical. Much of the genus looks too much alike to be identified by appearance alone, making accurate field identification difficult. They are occasionally mistaken for *Cortinarius* because of the cortina and brown spore print.

Some of the first gilled mushrooms to appear each year are *Agrocybe* species. They are saprobes that can be found on woody material, on manure, or in grassy areas. A common place to find them is in mulch or wood chips. *Agrocybe praecox* is a group of very closely related species that are commonly encountered. They have adapted well to the presence of human beings and seem to prefer areas where we have changed the environment. City parks, compost piles, and roadsides are the best places to look for them.

Agrocybes are edible, but not many are consumed due to their bitter taste. In some areas, this bitterness seems to be minimal and they are consumed. One species is a very popular cultivated mushroom in Italy known as pioppino. Some species of *Agrocybe* can be found in forest settings, growing from decaying wood or other debris. *Agrocybe* spores are dull brown, which makes them easy to distinguish from any toxic species of *Amanita*.

Part 2.4. Agarics with Dark Spores

The spore color of these mushrooms is much darker than that of any of the other agarics. They range from very dark brown to purple-black to totally black.

Part 2.4.1. *Agaricus* Species

Agaricus might be the most recognizable genus in North America; not from forest encounters, but from supermarket visits. *Agaricus bisporus* is often referred to as the "grocery store mushroom." White buttons, portobello, and crimini mushrooms are all *Agaricus bisporus.*

Aside from this domesticated species, *Agaricus* is a very large and complex genus. They are saprobes that can thrive in a wide variety of settings. Forests, pastures, and even deserts can be home to various species of *Agaricus.* All the mushrooms in *Agaricus* are medium-sized to large (2–6 inches [5–15 cm]) and generally white (some are brownish), have a ring on the stem, and have gills that become dark brown.

Although there are many nearly identical species, they can be divided into three major groups: Those that quickly turn yellow upon cutting or bruising, and usually smell of ink, carbolic, or phenolic; those that barely turn yellow or do not at all, and smell wonderfully sweet like almond or anise; and those that turn red (or brownish) upon cutting or bruising, and may smell mushroomy or slightly sour. As far as edibility goes, avoid rapidly yellowing, phenolic-smelling *Agaricus* species, as many are toxic. The two other groups have species prized for edibility.

One of the smaller, but still popular members of the genus is *Agaricus campestris.* Commonly known as the meadow mushroom, this pasture-dweller is common across North America and Europe. It is a prized edible that is very

Agaricus campestris

Agaricus arvensis

Agaricus cogwheel partial veil

Agaricus augustus (left) and *Agaricus moelleri* (right)

popular wherever it grows. The cap is usually white, but it can have brownish scales, and the mushroom will slowly turn reddish if sliced or bruised. The gills are free and pink when young, becoming dark chocolate brown as the spores mature. A membranous partial veil clings to the stalk as a cottony skirt-like ring after the cap has expanded. **Caution:** A deadly white *Amanita* species could be mistaken for the meadow mushroom, so use extreme caution when collecting, especially near ectomycorrhizal host trees. Check the gills and spore color!

Two very similar mushrooms in this genus are *Agaricus arvensis* and *A. sylvicola*. Some authors consider them varieties of the same species. *Agaricus arvensis* grows in grassy areas, while *A. sylvicola* grows in forest settings. Except for this difference in habitat and a slight difference in spore size, they are indistinguishable. Both are large, whitish mushrooms with grayish-pink gills that quickly darken to a deep, chocolate brown; both slowly bruise yellow. Both species are typically larger than *Agaricus campestris*, sometimes reaching a diameter of 10 inches (25 cm), and neither will have the characteristic pink gills that a young meadow mushroom has.

Possibly the most prized of all *Agaricus* species is the prince, *Agaricus augustus*. It is brownish, yellows slowly, and smells strongly of almonds. A look-alike (pictured with the prince) that can cause gastric upset is *Agaricus moelleri*. It is more grayish than brown, stains yellow rapidly, and has an awful phenolic smell. Both are common in the West and across Europe. All of the larger species of *Agaricus* have a persistent annulus that resembles a cogwheel, especially on the underside.

Keep in mind that while many species of *Agaricus* are edible, there are poisonous ones as well. The general rules are to avoid any *Agaricus* mushrooms that rapidly stain bright yellow when the base of the stem is cut and any that have a chemical smell when crushed. This odor is described as inky, phenolic, or resembling adhesive bandages. And one final note: Although *Agaricus* are urban mushrooms frequently collected for the table, even the edible species have some hazards. Do not pick them near roadsides or sources of pollution, nor from parks and lawns where chemicals are sprayed, or the like. Also, most *Agaricus* species contain potentially carcinogenic hydrazines (agaritine) and thus should be thoroughly cooked and eaten infrequently or in small amounts.

Part 2.4.2. *Stopharia* Species

Stropharia is a genus of thick-fleshed saprobes with purple-black spores. Though not common, they can be found across North America and Europe from early summer through fall, typically on wood chips or decaying wood. They are medium to large mushrooms (the caps are usually 2–4 inches [5–10 cm]) and most species boast a heavy, persistent annulus.

The type species, and one of the more colorful mushrooms in the genus, is *Stropharia aeruginosa*. It is a striking blue-green color when young, but it fades to yellow-brown. Although it is one of the smaller species of *Stropharia*, rarely exceeding 2 inches (5 cm) in diameter, a sighting is memorable. The cap is slimy when the mushroom is young, and the edge is often ornamented with partial veil remnants.

A larger and more familiar mushroom in the genus is the wine cap, *Stropharia rugosoannulata*. The wine cap is not often found in the wild, but it is widely cultivated in Europe and North America. It can be very large, sometimes greater than 10 inches (25 cm) across, hence its other common name, garden giant. It has a distinct annulus that resembles a cogwheel or a crown. The cap varies from pale brown to deep burgundy-red. Wine caps have a pleasant odor, often described as similar to asparagus or raw potatoes, and are easy to grow in great quantity. Their clean flavor and large size make them a very desirable mushroom.

Stropharia aeruginosa

Stropharia rugosoannulata

Part 2.4.3. *Psilocybe, Deconica,* and *Panaeolus*

Though these three genera are only distantly related, they do possess enough similar physical features to be grouped together here. *Psilocybe* is probably the most famous genus of mushrooms. This genus contains most of the hallucinogenic, or "magic," mushrooms. They are found worldwide, but the range of each species can be limited to very specific regions. They are all saprobes and most grow on wood chips or forest debris. *Psilocybe cyanescens* is one of the largest and most widespread in North America and Europe. A few species grow on manure, including *Psilocybe cubensis*, which is native to the Gulf Coast but more widely known from illegal cultivation these days. "Active" species contain psilocin and/or psilocybin, and they can be identified by the bluing reaction of the flesh when crushed and by their purple-brown spores.

Deconica coprophila is a very common little mushroom, typically found growing on clumps of manure. It is often mistaken for a species of *Psilocybe*, and some of the older guides have it listed as such. "Inactive" members of *Psilocybe* were moved to *Deconica* recently; they are not hallucinogenic and do not stain blue when crushed. The original type species for *Psilocybe* is *Deconica* (*Psilocybe*) *montana*, an associate of mosses in mountainous areas. Both small deconicas are widespread across North America and Europe.

One mushroom that goes unnoticed more often than not is *Panaeolus foenisecii*. This tiny mushroom is common in lawns across North America and Europe. It is usually too small to be seen during a casual walk through the yard and is very fragile. The cap is hygrophanous and has a darker band around the outer edge. It has many common names, including lawn mower's mushroom and the haymaker. Its dark, purple-black spore print and saprobic behavior occasionally gets it confused with *Psilocybe*.

Psilocybe cyanescens

Psilocybe cubensis

Deconica coprophila

Deconica montana

Deconica coprophila on dung

Panaeolus foenisecii

Part 2.4.4. *Hypholoma* and *Psathyrella*

Hypholoma is a small genus of purplish spored mushrooms that grow on wood. They are found in all types of forests in the northern hemisphere and can be quite common. *Hypholoma lateritium* is one of the most common, referred to as brick tops. It is found in large clusters on hardwood logs and stumps. The brick-red cap is smooth and usually remains convex, even after the mushroom is fully mature. The cobweb-like veil and light, purplish spore sprint are key characteristics. It is edible and pleasant-tasting while young, but make sure you don't confuse it with other mushrooms that may be poisonous.

Hypholoma fasciculare, also known as the sulfur tuft, is another common member of the genus. It gets its name from the color of the cap. The sulfur tuft is bitter tasting and has bright, sulfur-yellow gills that take on a greenish tinge as the mushroom ages. This is a poisonous species that, when young, could be mistaken for brick tops or velvet foot. *Hypholoma fasciculare* can be identified by its habit of growing on the decaying wood of conifers, but decaying wood can be difficult to identify accurately. Be careful!

Hypholoma lateritium

Psathyrella is a large genus and contains some of the most common mushrooms in the world. They are small to medium-sized mushrooms (1–2 inches [2.5–5 cm]) with fragile, hollow stems that "snap" when bent. They resemble ink caps (see page 56).

Of these common mushrooms, the most widespread in Europe and North America is *Psathyrella candolleana*. This mushroom is typically found on lawns, sprouting from dead roots or buried stumps. It might also be found in wooded areas, appearing on the ground, but growing from buried wood. This mushroom is extremely fragile and crumbles readily when handled. Like other *Psathyrella* species, this mushroom has dark brown spores, but the gills do not become inky.

Psathyrella echiniceps in the West and *P. delineata* of the East and Europe are common on rotting wood, and usually found in large clusters. Both are similar in appearance, with white stems (with no ring), and caps that may be fringed or rugose (wrinkled). The fragile stem and dark brown spore print are emblematic of the genus.

Lacrymaria lacrymabunda resembles *Psathyrella* enough to be included in this group. It is commonly referred to as the weeping widow. It is a more robust mushroom with a medium-brown cap that is densely covered in appressed (flattened) hairs. When young, the gills are mottled with darker and lighter shades of pinkish brown. Moisture droplets often form on the bottom edges of the gills. The dark spores then accumulate in the moisture droplets, giving them the appearance of black, teardrop-shaped exudate (hence the common name). The dusky spores also collect on the upper surface of the ring zone, leaving a characteristic dark band on the white upper stem. This mushroom is found in lawns and grassy areas.

Hypholoma fasciculare

Psathyrella echiniceps

Psathyrella candolleana

Psathyrella (Lacrymaria) lacrymabunda

Part 2.4.5. Inky Caps

The coprinoids are also known as the ink caps or the inky caps. The group is comprised of four genera: *Coprinus*, *Coprinellus*, *Coprinopsis*, and *Parasola*. Older books place them all in *Coprinus*.

The smallest and most fragile of these three groups is *Parasola*. The gills of *Parasola* species do not completely deliquesce (liquefy) as the other coprinoids do. They turn black and the mushroom withers away to nothing.

The most well-known coprinoid is the shaggy mane, *Coprinus comatus*. It is one of only a handful of species in the genus *Coprinus*. Shaggy manes are the largest of the coprinoids, sometimes reaching heights of 16 inches (40 cm) or more! The shaggy mane gets its name from the prominent scales on its cap, giving the mushroom a "shaggy" appearance. The stems are long, white, and hollow. If you look closely, you will see a cord in the center of the hollow stem. The annulus, if present, is fragile and low on the stem. This is one of several ink caps that can appear almost overnight. They have a complete life cycle of just a few days. The gills rapidly mature and then dissolve into an inky, slimy mess. The shaggy mane is found on lawns, compost piles, and rich humus. They are an excellent edible, but you have to get them early!

One major group of *Coprinellus* that can be found across North America and Europe is the mica cap group, *Coprinellus micaceus*. Mica caps can be found emerging from soil, leaf litter, or grass, but they actually grow from buried wood or dead roots. The cap color is tan to brown, and the stem is bright white. The mica cap is found in groups of anywhere from two or three to twenty or more. Like other ink caps, they are short-lived and dissolve into black "goo" in just a day or two. The name comes from the film of crystal-like granules (like flakes of mica) on the young caps. The granules tend to disappear as the mushroom ages.

The orange-mat coprinoid, *Coprinellus radians*, gets its name from the fuzzy mat of sterile, orange-brown mycelium that is produced on the woody substrate at the base of the cluster of mushrooms. Sometimes the orange mat (known as the ozonium) will be present, but the mushrooms will be absent. This mushroom is found on damp wood on the forest floor, and it might also be found growing on damp wood in basements. The stem is slender and white. The cap is conical to hemispheric, radially striate, and has surface granules, much like the mica cap.

Coprinopsis atramentaria is a species that produces coprine, a toxin that blocks the enzymatic breakdown of alcohol in the body (and buildup of acetaldehyde), causing extremely unpleasant gastric distress and a full-body burning sensation (hot flashes)—essentially a powerful hangover. Alcohol consumed up to 5 days after eating this mushroom could produce these symptoms. *Coprinopsis atramentaria*, also called the alcohol inky and tippler's bane, is one of the larger ink caps. It has a grayish (almost silvery) cap with a brownish tinge or flattened brownish scales near the center of the cap. The

Coprinus comatus

Coprinus comatus cross-section

Coprinopsis variegata

Coprinopsis atramentaria

Coprinellus micaceus

Coprinellus radians and ozonium

stems are pure white and fragile, with an inferior ring that is very low on the stem. It is found at the base of dead trees, stumps, or from dead, buried wood, and often in urban settings, as in the photo on page 57.

Arguably the prettiest of all inkies is the scaly ink cap, *Coprinopsis variegata*, a medium-sized ink cap, easily differentiated from others of the group by the large, prominent scales on its cap. It is also more brownish colored than similar-sized ink caps such as the shaggy mane and the alcohol inky. Although reported as edible, this mushroom can cause gastric upset and should therefore be avoided.

Part 2.5. Stalked Mushrooms with Teeth or Spines

These mushrooms produce fruit bodies with a stalk and cap, but spores are borne on teeth or spines that hang from the underside of the cap, rather than gills.

Part 2.5.1. Hedgehog Mushrooms and Hawk's Wings

Also members of the chanterelle family (*Cantharellaceae*), and like their chanterelle and black trumpet kin (page 21), the hedgehogs are highly prized as delicious edibles. They resemble chanterelles in overall shape and color when seen from above, but have spore-bearing spines or teeth below the cap. *Hydnum umbilicatum* is somewhat small (about 2–3 inches [5–8 cm] wide) and has sort of a "bellybutton" in the center of the cap; *H. repandum* can grow fairly large, with caps exceeding 6 inches (15 cm). Like chanterelles, all are mycorrhizal.

Also mycorrhizal (with spruce and fir) are the hawk's wings, *Sarcodon imbricatus*. These beautiful mushrooms can be difficult to see in dark forests as they blend in with surroundings. Look for their gray teeth underneath.

A very small stalked tooth mushroom that you are unlikely to see, but that is worth the search, is the earpick mushroom, *Auriscalpium vulgare*. This mushroom is common but rarely

Auriscalpium vulgare

encountered, as it grows solely on the cones of pine and Douglas-fir trees.

All the mushrooms in this section are common across North America and Europe.

Hydnum umbilicatum

Sarcodon imbricatus

Part 2.6. Astipitate Mushrooms with Teeth or Spines

These mushrooms produce spores that are borne on teeth or spines hanging from the underside of the cap, rather than gills, but the fruit body has no stalk.

Part 2.6.1. Lion's Mane and Similar

The lion's mane, or bearded tooth (*Hericium erinaceus*), is a large, stalkless tooth fungus found on broadleaf logs, stumps, and living trees. It is white and consists of one large clump of teeth. The basidia are borne on the outer surface of the teeth. This is one of the most commonly cultivated mushrooms in the northern hemisphere.

The bear's head tooth, *Hericium abietis*, is found on conifers, especially hemlock and fir. The similar *H. coralloides* is found on hardwoods. Both have numerous clumps of teeth, which are about ½ to ¾ inch (1–2 cm) long. All range across North America and Europe. The comb tooth (*Hericium americanum*) has relatively short teeth (approximately ¼ inch [0.5 cm] long), which come off from both sides of the branches

Hericium erinaceus

Hericium abietis

Climacodon septentrionale

and, unlike the teeth of other hericiums, point in all directions. All look like pieces of white coral, and all are tasty edibles.

The northern tooth, *Climacodon septentrionale*, is a stalkless toothed mushroom that grows on standing trees (mostly maples), with overlapping shelves often seen very high above ground. The teeth are much tougher than other fleshy tooth fungi (*Hericium* species). This mushroom can grow quite massively, often leaving a light area on the bark where the fruit body grew the previous year. It is not edible.

Part 2.7. Mushrooms with Stalk, Cap, and Pores

A bolete is a fleshy mushroom that lacks gills but has a layer of tubes instead. These tubes are arranged in an often-separable layer, ending in a surface of pores from which spores are released into the air. Boletes make up an enormous group of macrofungi; many dozen genera and hundreds of species inhabit every forest in the world. We will break down a few of the more commonly encountered groups, based on genetic relation or just on physical similarities.

Part 2.7.1. King Boletes

One of the finest of edible mushrooms is the king bolete. It goes by dozens of common names around the world and is actually a group of many similar species. *Boletus edulis* of Europe is the most widely known, but there are at least a dozen others that are nearly identical in taste, texture, and appearance.

All species within this group share a set of common physical features. They are large mushrooms, sometimes growing to weights of 3 pounds (1.5 kg) and reaching diameters of 15 inches (38 cm). They all have white pores when young. Their pore surfaces are compact and look cottony. As these mushrooms mature, the pores will start to turn an olive yellow-green, never pinkish. Near the top of the stem, we will see fine reticulation. (Reticulation is a net-like pattern of raised tissue.) The highest lines of the reticulation will be lighter colored than the background in this group of boletes. No boletes

Boletus edulis, the king bolete, Italy

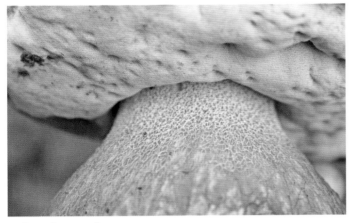

Boletus barrowsii

Boletus rubriceps

in this clade will stain or bruise when cut. They are very firm, especially when young, and keep their texture when cooked gently.

The king boletes of North America range from almost white, such as *Boletus barrowsii*, to tan and brown, to reddish-brown, such as *B. rubriceps* of the Rocky Mountain region. No matter the cap color, all feature a fat stem (often wider than the cap) with white reticulation near the top.

Part 2.7.2. Scaber Stalks

Leccinum is one of the largest genera within the bolete family. They are mycorrhizal, as almost all boletes are, but they are very host-specific. Identifying the tree, in most cases, will help to identify the mushroom. All species of *Leccinum* have stems that are heavily decorated with "scabers," rough protrusions that cover the stem. In most species, these scabers are much darker than the stem, making them quite visible. In a few species, they may be so pale that they are easier felt than seen. Upon close examination, these bumps are softer than their name suggests. Another distinguishing feature of some species is a flap of sterile tissue along the edge of the cap. This can be very important when trying to identify a *Leccinum* to species.

A common species across Europe and northern North America is *Leccinum scabrum*. It is found exclusively with birch trees (*Betula* species), which is where it gets its common name, brown birch bolete. The cap is dull brown when young, but cracks with age, revealing the nearly white flesh beneath. The pore surface is pale gray and bruises brown when handled. This mushroom is edible but be certain of your identification.

Leccinum scabrum

Leccinum margin

Other leccinums occur with aspen trees or oaks. *Leccinum vulpinum*, or scaber stalk, is a striking orange-red mushroom common across northern North America and Europe with pines and spruce. Some *Leccinum* species have been known to make people sick, especially when eaten raw.

Leccinum vulpinum

Part 2.7.3. Slippery Jacks

The slippery jack and its relatives make up the genus *Suillus*. They can be found with conifers across the northern half of the globe. Most are small- to medium-sized mushrooms, rarely exceeding 6 inches (15 cm) in diameter. They get their common name from having a slimy or sticky feel to the cap.

Not all species have this gelatinous layer, however. *Suillus spraguei* of eastern North America (*S. lakei* of the West is similar) has a scaly, dry cap. It is often found in bogs and is associated mostly with eastern white pine (*Pinus strobus*). This mushroom gets the nickname painted bolete for its brilliantly colored features. The cap and stem are red and pink with yellow highlights, and the pore surface is bright yellow but may be hidden beneath a partial veil when immature. Although it is edible, it has an odd texture. This mushroom is better photographed than eaten.

Similar but more brownish and furrier is *Suillus cavipes*, which is noted by its hollow stem. This mushroom, along with the slimy red-brown *Suillus grevillei*, occurs with larch (tamarack) trees across northern North America and Europe. The mushroom most commonly referred to as the slippery jack, *Suillus luteus*, is common across North America and Europe under pine trees. This small brownish bolete, to about 5 inches (13 cm) across, has a very slimy membrane covering the cap, especially when young. The yellow pores are protected by a partial veil that leaves a purplish-brown ring on the stem of the mature mushroom. This mushroom is collected for the table more than any other in the genus. Similar is the chicken fat mushroom, *Suillus americanus* (*S. sibericus* in the West), which is slimy and yellow and occurs with white pine.

Suillus (pictus) spraguei

Suillus grevillei

Suillus luteus

Suillus cavipes

Suillus americanus

Part 2.7.4. Bitter Boletes

Mushrooms in the genus *Tylopilus* are often mistaken for good edible boletes, but one bite will change your mind. They are not poisonous, but they are extremely bitter, and therefore considered inedible.

Tylopilus felleus is a common mushroom in North America and Europe. It does have a strong physical resemblance to members of the king bolete group, but it is definitely not as tasty. Looking closely at the reticulation on the upper stem, you will see the highest edge of the pattern is darker than the background, exactly the opposite of prized king boletes. This mushroom is found growing with conifers and can occasionally be found growing from very decayed stumps. The pore surface is white when young but turns pink with age and bruises dark pink when handled.

A very large and very common species from North America is *Tylopilus rubrobrunneus*. This reddish-brown mushroom is much larger than most other boletes, often growing to a diameter of 12 inches (30 cm) or more. Even when fully mature, the cap tends to remain convex. Though of tempting size, this mushroom is far too bitter to be considered edible.

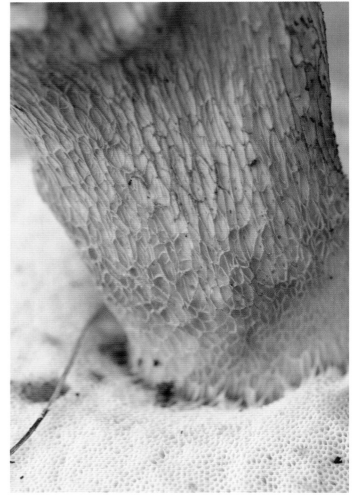

Tylopilus felleus close-up

Tylopilus rubrobrunneus close-up

Part 2.7.5. Two-Colored Boletes

The two-colored boletes are some of the most difficult to identify to species, sometimes even to genus. This group of mushrooms is enormous and there are dozens of species that could be called two-colored boletes. All these mushrooms have some shade of red on the cap and yellow pores. The stem can be red or yellow and often has both colors present to varying degrees.

Baorangia (*Boletus*) *bicolor* is the most widely known of the two-colored boletes, but it is probably misidentified as often as it is identified correctly. It is a species complex that can be found in North America and Europe. It is one of the larger boletes in this group, sometimes reaching a diameter of 8 inches (20 cm). The bright yellow pore surface bruises dark blue when handled, but the flesh does not, or sometimes does only faintly. It has a pleasant odor and is a great edible, but accurate identification is key. Others that look similar may cause sickness.

If you are not proficient with bolete identification, you have to be careful of some that bruise blue. Indeed, a safe bet in North America is to avoid blue-staining boletes with red pores. We have two deadly species that occur in the

Boletus bicolor

Boletus eastwoodiae

Boletus satanas, Italy

West and are known as Satan's bolete, *Rubriboletus* (*Boletus*) *eastwoodiae* and *R. pulcherrimus*. Both have lightly colored caps and a striking red stalk and pores. The "true" Satan's bolete of Europe is *Boletus satanas* and is similar, but the cap is white.

Part 2.7.6. Ornamented Boletes

Some boletes are wildly decorated with exaggerated features. They are not closely related but are grouped together here to aid in identification. Some have stem reticulation that you can get lost in, while others have tufts on the cap that can resemble the armor of a prehistoric beast.

One famous species of ornamented bolete is *Strobilomyces floccopus*. Members of *Strobilomyces* are often called old man of the woods. They are gray mushrooms with black decorations on the cap. They have a very cottony partial veil that protects the developing gills. This often leaves a ragged edge on the mature cap. The flesh of some species of *Strobilomyces* stains an odd, reddish color when exposed to air. The red becomes darker until it eventually appears black. These beautiful mushrooms can be found in almost any forest in the northern hemisphere.

A mushroom with exaggerated reticulation of the stem is Russell's bolete, *Aureoboletus* (*Boletellus*) *russellii*. The texture of the stem is garish, boasting ridges and valleys that are surprisingly large. This striking mushroom is abundant during some years and completely absent in others. It is most common in eastern North America, but it may be present in eastern Europe. The cap is relatively small in comparison to the height of this bolete. The stem and cap are brown and the pores are bright yellow, giving great contrast.

There are many other boletes that can be considered ornamented as well.

Strobilomyces floccopus

Boletellus russellii

Part 2.7.7. Stalked Polypores

If you check the spore-producing layer of a mushroom and see pores or tubes (rather than gills or teeth), you can be pretty sure that you are looking at a bolete or polypore. Boletes look like agarics but with tubes. Polypores have many pores, hence the name, and usually grow in a shelf-like manner from wood. However, there are a few stalked polypores, and they are pretty easy to identify. Most grow from wood, live or dead, and in some cases the wood may be buried.

Lentinus (Polyporus) brumalis, winter polypore, is a small, stalked polypore with elongated white pores. When young, the 2-inch (5 cm)–wide brown fruit bodies are velvety to hairy, but at maturity they are smooth or slightly velvety and gray-brown. *Lentinus (Polyporus) arcularius*, fringed polypore or spring polypore, is similar-looking, but the brown cap is scaly and the pores are large, angular, and honeycomb-like. The most notable feature of the fruit bodies is the hairy cap edge. Both have central stalks and appear on dead hardwoods (wood might be buried) in late summer through fall, but they can be found all winter long. Both are known in eastern North America and Europe.

Picipes (Polyporus) badius (see page 70), big black-footed polypore, is a medium-sized (8 inches [20 cm] across), stalked polypore with a distinctly dark lower stem. The stalk is usually off-center to lateral, although some specimens might have a central stem. The cap surface is pale, becoming dark orange-brown. Check

Polyporus badius

Polyporus elegans

Neofavolus alveolaris

that stalk again; the pores of *P. badius* extend somewhat downward on the stem and contrast sharply with the dark stem.

This mushroom is similar to the smaller species the little black-footed polypore, *Cerioporus* (*Polyporus*) *varius*, which is another stalked polypore with a black lower stem. The elegant polypore (*Polyporus elegans*) is nearly identical; both mushroom's stalks can be central, eccentric, or lateral. The cap (4 inches [10 cm] across) has fine white pores below; the top is usually a light tan, helping to distinguish this species from the large black-footed poly-pore, *Polyporus radicatus*, which has a cap that is usually dark orange-brown. All three are known across North America and Europe.

Neofavolus (*Favolus*) *alveolaris*, the hexagonal-pored polypore, is a small (to about 2 inches

Polyporus squamosus

Phaeolus schweinitzii

Polyporus radicatus

[5 cm] across), annual polypore that occurs on sticks and small logs. It has a reddish-brown cap and a white to cream pore surface. The pores are large and angular, and as the name implies, many are six-sided. The largest pores are typically in the center with smaller pores around the margin.

Cerioporus (*Polyporus*) *squamosus* (see page 71), dryad's saddle or pheasant's back, is found on dead or dying trees or on fallen logs. It is particularly common in spring at the base of dead elms. It has a lateral stalk that is dark brown to black, and it has large, angular pores and brown scales on the cap. This very large polypore can be more than 12 inches (30 cm) across and is hard to miss during morel season in eastern North America and across Europe.

Phaeolus schweinitzii (see page 71), dyer's polypore, is a serious pathogen of conifers across North America and Europe. It causes a brown, cubical rot in the large basal roots and lower portion of the tree's trunk. The fruiting bodies can be very large (more than 12 inches [30 cm] across) and usually are found growing from roots near the base of the tree in overlapping rosettes of shelves. These mushrooms can be very photogenic; when young they are brown with yellowish edges and a velvety surface. As they age, the velvety surface disappears and the whole fruiting body becomes dark brown. This mushroom is quite common in the fall wherever conifers (especially five-needle pines) are found, but the fruit bodies may dry out and persist into the following year. Fresh specimens are highly prized by those who dye fabrics using natural dyes.

Part 2.8. Mushrooms with Stalk and Wrinkled, Cone-shaped, or Saddle-shaped Cap

This section includes mushrooms that are somewhat similar in appearance, even though they are separated by millions of years of evolution. Some, such as the stinkhorns, are basidiomycetes, while others, such as the morels, are ascomycetes.

Part 2.8.1. Morels

There are three morphological groups of morels: Black, yellow, and half-free. Historically, they were thought to be three species of morels, but it has been determined that there are many species in each of these three groups and all but a few look identical.

One thing is certain: Morels are the most popular of wild edible mushrooms. And for most who pursue these ascomycete mushrooms, it's most useful simply to refer to them as black morels, yellow morels, and half-free morels. The difference is based not so much on color as it is on the shape of the convoluted cap. With yellow morels, the bottom of the cap is attached directly to the stem, black morels have a loose overhang (termed *vallecula*) at the bottom of the cap, and half-free morels have the bottom half of the cap hanging down, free from attachment with the stem.

Other facts to know: All morels can be gray in color when immature, and true morels are

Morchella snyderi

entirely hollow. (Be sure to note differences between true morels and false morels; see page 57).

Black morels (*Morchella snyderi*, *M. elata*, *M. angusticeps*, and others) are typically 1–2 inches (2.5–5 cm) wide and tall. This mushroom may occur in large groups over a large area. Look for it in woods of deciduous or conifer trees, often under tulip trees and ash trees. Black morels may be associated with forest fires.

Yellow morels (*Morchella americana*, *M. esculenta*, and others) are typically 2–3 inches (5–8 cm) wide and 3–8 inches (8–20 cm) tall. This mushroom is found singly or in clusters on the ground and typically with deciduous trees, including dead elm trees in eastern North America, tulip trees, and ash trees, and in old apple orchards.

Half-free morels (*Morchella punctipes*, *M. populiphila*, and *M. semilibera*) are much smaller and thinner than other morels, but much the same color as yellow morels. This mushroom is elongated and narrowly conical with about

Morchella esculenta

Morchella punctipes

half of the cap hanging free below the point of attachment with the stalk. The cap can be disproportionately small compared to the stem, which continues to grow into maturity (up to 6 inches [15 cm]). Look for this mushroom in deciduous woods, often under tulip trees, ash, poplars, and cottonwoods. Due to its smaller size, this mushroom, although edible, is not as highly prized as other morel species. What it lacks in size, however, this morel makes up for in numbers—hundreds can be found at a time.

Half-free morels could be confused with verpas; a species with a smooth cap is *Verpa conica*, while *V. bohemica* has a convoluted cap. Both are close relatives of true morels but distinguished by having a hollow stem stuffed with cottony fuzz; they are known from across all of North America and Europe.

Verpa conica

Part 2.8.2. False Morels

Gyromitra species are known as false morels. They co-occur with true morels in the spring and bear a slight resemblance to them, but they have several obvious differences. False morels do not have one large, hollow cavity from top to bottom as morels do, and false morels have a wavy appearance as opposed to the deeply pitted caps of true morels.

That many *Gyromitra* species look somewhat similar to *Morchella* species is due to the fact that they are closely related. Most books list all false morels as toxic and warn against eating them. In truth, they are commonly collected for the table, and many in North America and Europe regard them as tastier than true morels! Based on scientific analysis, it seems that *G. infula* (which occurs on rotting wood) and North American *G. esculenta* may be the only species with toxins—hydrazine compounds similar to those in rocket fuel. These species should be avoided. All others, if eaten, should be fully cooked, as any traces of volatile hydrazines are driven off with heat.

The largest of the false morels is the red morel, *Gyromitra caroliniana* (also called "big reds"), which may exceed 8 inches (20 cm) and have enormous stems. Big reds are found in mixed forests in eastern North America, as are pig's ears (also called gabled false morels), *Gyromitra brunnea*. The latter is a stocky false morel (growing to about 5 inches [13 cm]), but more brownish, with the appearance of a tri-cornered hat.

The conifer false morel, *Gyromitra esculenta*, is common across northern North America and Europe. Reddish to very dark with a slender stalk, it grows to about 4 inches (10 cm). The bull-nose false morel is also seen across North America and Europe, but typically with hardwoods. This curious little mushroom is

Gyromitra caroliniana

Gyromitra brunnea

Gyromitra esculenta

Gyromitra korfii

Gyromitra brunnea **cross section**

short and stocky (growing to about 4 inches [10 cm]), with a cap margin that hangs down to the ground, hiding the stem. Curiouser still is that scientists cannot decide what to call it; this mushroom goes by *Gyromitra korfii* in the East and *G. montana* in western North America, and additional names in North America and Europe are *G. gigas* and *G. fastigiata*. In time, it may be determined that these are all the same species or several different ones.

Part 2.8.3. Elfin Saddles

Helvella species are ascomycetes and are fairly closely related to *Morchella* and *Gyromitra*. Sometimes they are called false morels, but they are better known as saddle fungi or elfin saddles for the overall morphology of most species. Some, such as *Helvella lacunosa*, actually resemble gyromitras; one *Gyromitra* (*G. infula*) closely resembles a saddle-shaped *Helvella*.

Helvella species mostly fruit in the summer through fall. They can be found in a variety of habitats including both mixed forest and disturbed areas, along paths and grassy areas, and in compacted and even gravely soils.

Helvella crispa

Helvella vespertina family

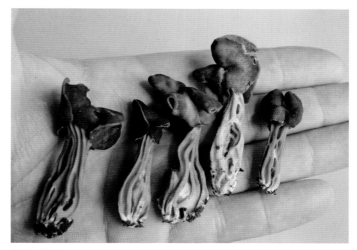

Helvella lacunosa

Until recently, all helvellas with a fluted stalk and gray to black convoluted cap were called *Helvella lacunosa*, a common species known across Europe. It has recently been determined that it probably does not occur in western North America; for eastern North America, the jury is still out but the name continues to be used.

Helvella vespertina can grow to 6 inches (15 cm) or more, has a very convoluted (lacunose) stalk, and has a gray to black convoluted cap. It is common in the West, and it is the largest *Helvella*; *H. lacunosa* and *H. dryophila* are similar species but rarely exceed 4 inches (10 cm). The fluted white helvella (*H. crispa*) is similar but with a white cap; it is found across North America and Europe.

There are also many helvellas with smooth stalks. Probably the most common from across North America and Europe is *Helvella elastica*, which also grows to about 4 inches (10 cm).

Helvella elastica

Part 2.8.4. Stinkhorns

Stinkhorns bear a slight resemblance to morels, but they are basidiomycete mushrooms like puffballs, gilled mushrooms, and boletes. All stinkhorns arise from a cuplike volva and produce a slimy, foul-smelling mass of spores on the cap. All stinkhorns arise from an "egg," an immature stage. Slice open the egg to ensure that it is not confused with an edible puffball or the immature egg of an *Amanita* mushroom.

Each of these large stinkhorns has a white stalk and a grayish-green mass of slime at its tip. The foul-smelling mass of spores attracts flies, which pick up the spores on their legs and bodies and carry them to distant locations. *Phallus impudicus* and *P. hadriani* have convoluted caps, somewhat like a morel, grow to 6 inches (15 cm) or more, and range across North America and Europe. *Phallus ravenelii* is similar, but with a smooth cap and purple rhizomorphs attached to immature egg.

The elegant dog stinkhorn (or devil's dipstick), *Mutinus elegans*, has no cap; the slimy gleba is borne at the tip of the orange to pinkish-red stalk. The dog stinkhorn, *Mutinus caninus*, is very similar but with a whitish stalk. Both grow to 6 inches (15 cm) and range across North America and Europe.

Slenderer still, and with a small cap, is *Phallus rubicundus*, which also is found across North America and Europe. Stinkhorns come in many other bizarre forms, hardly resembling mushrooms at all. One that looks like a garish wiffleball, commonly seen in the West and Southeast and in much of southern Europe, is the basket stinkhorn, *Clathrus ruber*. Despite their off-putting smell, stinkhorns are consumed while in the egg stage by some adventurous mycophiles.

Phallus impudicus

Mutinus elegans

Clathrus ruber

Phallus ravenellii

Phallus ravenellii with eggs cross-sectioned

Phallus rubicundus

Part 2.9. Shelf-like Mushrooms with Pores

Polypore is a general term for mushrooms whose fruit body is a single shelf or group of shelves with pores on the underside.

Part 2.9.1. Perennial Polypores

Perennial polypores persist on their host tree for multiple years, putting a fresh new hymenial layer on the underside each growing season. They are typically very hard and woody, and if sliced in half they even display growth rings.

The artist's conk, *Ganoderma applanatum*, is one of the most common bracket fungi found across North America and Europe. It gets its name from the fact that the fresh pore surface can be easily engraved with a knife, sharp stick, or fingernail. The conks are perennial and put down a new layer of tubes each year. Young conks might be less than an inch (2.5 cm) in diameter, while old specimens might be 2 feet (61 cm) across or more. The conks vary in color from ash gray to tan to brown. This fungus may be found on hardwood and conifers. It is primarily a rotter of dead trees and downed logs, but it will occasionally cause decay in living trees. Copious amounts of brown spores may be deposited on any surfaces nearby. This species has become popular for its purported medicinal properties.

Phellinus tremulae (called horse hoof fungus) is one of the most common perennial conks of North America and Europe, where it is found on aspen trees (*Populus* species). There are several other horse hoof fungi, all fairly host-specific. They are recognized by having a hard, wavy upper surface and layers that stack up over the years to resemble a horse hoof. The pore surface is brown and velvety with very small pores. A large conk of *Phellinus* species could be more than 30 years old, making this one of the longest-living fruiting bodies in the fungal kingdom.

Other common species of this genus are *Phellinus igniarius*, the tinder polypore, which grows on birches, aspens, willows, and cottonwoods, and *P. rimosus*, which is found on black locust. *Phellinus gilvus* (mustard yellow polypore) will have a yellow color on the edge of the cap of very fresh specimens. More often, the color is actually a muddy brown on both the top and bottom surfaces, and the interior often has a rusty or dark mustard color, which can be seen if the conks are broken apart. *Phellinus gilvus* causes a decay of the sapwood of hardwoods and is widespread; *Fomitopsis pinicola*, the red-belted polypore, is exceedingly common on conifers of the northern hemisphere.

Another perennial polypore to know is *Heterobasidion annosum*, the conifer-base polypore. This fungus is very photogenic but extremely destructive, rotting the heartwood of conifers. It occurs across North America and Europe, where it is often overlooked if the conks are not well developed or if they are hidden among debris and moss.

Ganoderma applanatum with brown spore deposits

Ganoderma applanatum, the artist's conk, being drawn on

Phellinus tremulae

Heterobasidion annosum

Fomitopsis pinicola

Part 2.9.2. Annual Polypores

Annual polypores typically sporulate one year and then die, although they may persist for a period of time before falling off their host tree. They are usually not hard or woody, but more rubbery and pliant. Some are even collected as edible mushrooms.

Ganoderma tsugae, the varnished conk, is found almost exclusively on hemlock trees in eastern North America, while the similar *G. oregonense* is found on conifers in the West. *Ganoderma lucidum* (ling chih) is also similar and occurs on hardwoods as well as some conifers; it ranges across North America and Europe. The cap may be a foot (30 cm) across or more, and it is colored a deep reddish-brown to dark maroon; the growing edge is white, as is the underside. When fresh, these mushrooms are very shiny and appear to be coated with varnish. Ling chih has long been used for its purported medicinal properties.

Ischnoderma resinosum, resinous polypore, is soft and almost rubbery when fresh and has a thick, rounded margin (resembling a "fat lip"). A close examination reveals shallow, radial furrows. This mushroom grows to about 9 inches (23 cm) and exudes water droplets as it ages. The water droplets pick up some of the pigment of the conk and look like resin, thus the name. Although it is bland, some people collect this mushroom as an edible.

Tyromyces chioneus (cheese polypore) is soft and spongy (if it is squeezed, water may drip from its fruit bodies), and white to grayish. *Tyromyces caesius* (*Postia caesia*) is similar but with bluish-gray shades on the surface of the cap, hence the name blue cheese polypore. Both species grow to about 3 inches (8 cm) across, are found widespread throughout North America and Europe, and despite the names, are not edible.

Although few polypores are collected as edibles, one very noteworthy "polypore" is quite popular in Europe (less so in North America): *Fistulina hepatica*, the beefsteak polypore. This dull red polypore grows from the wood of chestnut trees and their relatives, including some oaks and chinquapin. When fresh, the mushroom is glutinous and looks every bit like a piece of raw meat or liver, especially when in cross section. This enigmatic mushroom has the growth habit of a polypore, but its tubes are not fused together as a solid fruit body like true polypores' are, and indeed it is actually more closely related to gilled mushrooms.

Many *Ganoderma tsugae* on a tree

Ganoderma tsugae close-up

Ischnoderma resinosum

Tyromyces chioneus

Fistulina hepatica on wood

Fistulina hepatica cross section

Part 2.9.3. Polypores with Overlapping Caps

Grifola frondosa, or hen of the woods, consists of many spoon-shaped fleshy outgrowths clustered around a large, branching stalk. The pores extend downward on each of the secondary stalks. "Hens" are also known as sheepshead in some places. They are typically found at the base of oak trees, but they can be associated with other hardwoods. They are common in eastern North America and Europe and rare in western North America. The color varies from whitish to gray to red-brown. Some of the larger clumps of this species have weighed more than 100 pounds (45 kg)! This is considered a choice edible, but it must be collected when fresh.

Grifola frondosa

Meripilus giganteus

Laetiporus sulphureus

Laetiporus sulphureus and related species are known as sulfur shelf and chicken of the woods. These lovely, tender polypores vary from a bright sulfur-yellow to coral to orange. On the underside, the pore surface is lemon-yellow (the pore surface of *Laetiporus cincinnatus* is white). As the fungus dries, the color fades to white and the edges become brown. *Laetiporus sulphureus* is common in the East and throughout Europe on hardwoods; *L. cincinnatus* is less common (and is more peach-colored and fruits in rosettes, often from buried wood). In the West, we have *L. gilbertsonii* on hardwoods (especially *Eucalyptus* species) and *L. conifericola* on conifers. All are delicious and popular edibles, but they must be collected when young and tender; as they mature, the texture becomes poor and they may be riddled with insects. The flesh taste and texture are remarkably similar to that of the white meat of chicken.

There are some similar polypores that fruit in clusters at the bases of dead trees, including *Meripilus giganteus* and *Bondarzewia mesenterica* in Europe. In North America, *B. berkeleyi* (Berkeley's polypore) is found in the East and *B. montana* is found in the West. All can be very large, with clusters more than one foot (30 cm) across, and have caps in shades of brown and tan. *Meripilus sumstinei* of eastern North America is more grayish. *Bondarzewia* species can be identified by their spiny spores. (Anton Bondartsev was the legendary polypore specialist of Russia; the Reverend Miles Berkeley was one of the "founding fathers" of mycology.) *Bondarzewia* and *Meripilus* species are edible but bland.

Part 2.9.4. Turkey Tail and Similar

There are many species of small polypores that grow on decaying wood. One that grabs the attention of foragers around the world is *Trametes versicolor*, commonly called turkey tail. This mushroom is often collected for its purported medicinal value. Turkey tail is widespread and common throughout the northern hemisphere. It is found on the dead branches and decaying stumps of various hardwoods. It looks like small, leathery flaps, sometimes growing in large clusters. The upper surface varies greatly in color, but always has a basic brown element. There can be concentric bands that grow in different layers of color, much like the rings of trees. They can be yellow, purple, or even orange throughout the layers of the surface. The outer edge is typically white. The underside is pale gray to white, and the pores are very small; sometimes not visible to the naked eye.

Stereum ostrea, false turkey tail, is often confused with turkey tail. This similar-looking structure shares the color and shape of turkey tail but is lacking one major feature: Pores. *Stereum ostrea* is not a polypore, but rather a crust fungus. The flaps are very colorful, similar to turkey tail, but often with the addition of green from algae that commonly chooses to colonize this fungus. The flesh of false turkey tail is much thinner and more papery than that of turkey tail, and the underside has a slight, brownish tinge. Young specimens may be hairy, but this disappears with age.

Trametes versicolor

Stereum ostrea

Cerrena unicolor **close-up furry top**

Trametes (Pycnoporus) cinnabarina

Another small bracket that is common worldwide is *Trichaptum biforme.* It is more commonly referred to as the violet-toothed polypore. The common name is misleading; this is a polypore, not a tooth fungus. The individual pores separate, giving the appearance of teeth. This "toothy" surface is a beautiful, pale purple when the mushroom is young. This mushroom can be found across North America and Europe, and it is extremely common in the East. It can be found on many different hardwoods, but it shows preference to birch (*Betula* species).

Trametes (*Pycnoporus*) *cinnabarina* is one of the most striking and easily identifiable of the small, annual conks. It has an overall red-

Trichaptum biforme

Cerrena unicolor

orange color and a smooth to rugose (wrinkled), dry cap. The underside of the cinnabar red polypore is a more intense version of the cap color. This kidney-shaped mushroom has the texture of a dry sponge. It prefers the wood of fruit trees and often grows in open, sunny areas. This mushroom is valuable as a source of dye for fabric and yarn.

Mossy maze polypore, *Cerrena unicolor*, is a small shelf fungus with a densely hairy upper surface that often accumulates masses of green algae, giving it the appearance of being mossy. In many specimens, however, the algae are absent and the cap color is gray and faintly zonate. The lower surface is labyrinthine (maze-like) and is almost always smoky gray in color. The name translates to "a fungus of one color." This boring name is fitting for this mushroom. It can blend in with its hardwood host as a background and become nearly invisible. As the undersurface matures, it will appear more like teeth than a maze pattern.

Cerrena unicolor **close-up maze underside**

Schizophullum commune

Part 2.10. Shelf-like Mushrooms with Gills

At first glance, these shelf-like mushrooms growing from wood may appear to be polypores, but they have gills on their undersurfaces.

Part 2.10.1. Shelf Mushrooms with Split Gills or Maze Gills

These mushrooms grow in a shelf-like manner on wood, but underneath, rather than a surface of numerous small pores, the surface is comprised of expanded pores that resemble ribs or gill-like structures. All species featured here are cosmopolitan and common across North America and Europe.

Schizophyllum commune, the split gill, is aptly named. This small, white to gray mushroom grows on wood (particularly on small branches and sticks) on the forest floor. The gills (when young) and cap are hairy. This is particularly obvious at the margin of the cap, which has a fringe of dense hairs. The common name refers to the characteristic gills (folds), which are split lengthwise. This fungus has been widely used in studies on fungus physiology. And although small (2 inches [5 cm]) and very leathery, it is collected as an edible mushroom in Southeast Asia.

The oak maze polypore, *Daedalea quercina*, is large shelf or conk (8 inches [20 cm] wide) with a labyrinthine (maze-like) pore surface and beautiful shades of brown on top. (In Greek mythology, Daedalus was the inventor who created the labyrinth for King Minos.) This species is found on logs and stumps, usually of oak. *Daedaleopsis confragosa* and *D. septentrionalis* are similar to the above, but less thick and with thinner gills; both rot a variety of woods. All three are quite common.

Daedalea quercina

One of the most common polypores in the northern hemisphere is the turkey tail, *Trametes versicolor*, a small, thin shelf (see page 85). But some *Trametes* species can be larger and thicker. The birch mazegill, *Trametes betulina*, has velvety pale to light brown zonate (striped in a semicircular fashion) caps that overlap, and it is found on hardwoods. This species is also known as *Lenzites betula*.

Trametes elegans, the ambiguous polypore (older books may call it *Daedalea ambigua*), and *Trametes gibbosa* form thicker brackets that can be up to 12 inches (30 cm) across. Both form solid, tough (not leathery) conks with an upper surface that is typically white to gray and minutely velvety.

Part 2.10.2. Oyster Mushrooms

Oyster mushrooms occur all across North America and Europe. They are very popular edibles—so much so that they're cultivated worldwide—and pretty easy for beginners to learn. Note that there are some look-alikes ("imposters," see page 90) to be aware of.

Most oyster mushrooms grow in an overlapping, shelf-like manner on fallen logs and dead trees of many species; however, they're most commonly found on hardwoods. These are typically fairly thick-fleshed, large (the caps can exceed 6 inches [15 cm]) mushrooms. The stalk is lateral (not central) in most species and is sometimes nearly absent. Color is highly variable and ranges from white to yellowish to gray-brown. The gills are always white; the spore print is always white to very light gray.

The common name oyster mushroom refers to its shape (which resembles an oyster shell), not its flavor. Oyster mushrooms have a wonderful mushroomy, fruity, or anise scent when fresh; during the first instant of cooking, they may release a wonderful seafood aroma.

Pleurotus ostreatus

Pleurotus ostreatus

Hypsizygus ulmarius

Pleurotus dryinus

Pleurotus ostreatus may be found through much of the year and even into winter; it is light to somewhat dark gray-brown. *Pleurotus populinus*, the aspen oyster, is very commonly found as it fruits right after morels in late spring and into summer; it is restricted to aspen and cottonwood trees (*Populus* spp.) and has an anise smell. *Pleurotus pulmonarius*, the summer oyster, is very similar to the aspen oyster with light coloration but without the anise scent.

Hypsizygus ulmarius (formerly known as *Pleurotus ulmarius*), the elm oyster, is common on box elder (*Acer negundo*) and some other hardwoods. It is edible but not choice; when mature it can be very large and tough. It is recognizable by its host tree and central stem. A similar-looking oyster with a central stem and remnants of a veil covering over the gills (often there is only a slight ring left behind on the stem) is *Pleurotus dryinus*, the veiled oyster. It is less common, is often found on maple and other hardwood trees, has a furry stem when mature, also gets very tough, and is not sought out as an edible.

Part 2.10.3. Oyster Mushroom Imposters

There are several kinds of mushrooms that resemble oyster mushrooms. Each has a characteristic to help distinguish it from a true oyster mushroom, however. Although they are pretty to look at, none should be considered edible, and indeed, some of these imposters are poisonous. All the species shown here range across North America and Europe.

Pleurocybella porrigens, known as angel's wings, occurs on conifer wood and has a white spore print, but is usually smaller (caps are about 3 inches [8 cm]) than *Pleurotus* species, with flesh much thinner and flimsier. This mushroom should be avoided, as it has been implicated in poisonings. *Crepidotus* species,

Pleurocybella porrigens

Crepidotus mollis

Panellus stipticus

Phyllotopsis nidulans

known as creps and oysterlings, should also be avoided, as some are known to be poisonous. *Crepidotus* species are entirely stalkless and have gills that turn brown, denoting a brown spore color. They are usually pretty small, about 2 inches (5 cm) or less, and white to tan in color.

The orange mock oyster, *Phyllotopsis nidulans*, is one of the easiest of all fungi to identify, provided the specimens are fresh. The mushroom has a bright yellow-orange color, a densely hairy cap that grows to about 3 inches (8 cm), and an incurved margin (even when

Panus conchatus

mature). And although it is very photogenic, this mushroom has an extremely foul sour smell! The orange mock oyster is found in overlapping clusters on deciduous wood in late fall and sometimes in early spring. It is not a true oyster mushroom, but the violet oyster, *Panus conchatus*, does look similar; however, it is quite furry and has beautiful violet tones. It is widespread on wood but not collected as an edible.

The luminous panus, *Panellus stipticus*, is a small (1-inch [2.5 cm]), oyster-like mushroom often found in clusters on stumps and logs, especially oaks. With a hand lens, you can see cross veins between the forking gills. It has a tan fleshy cap with a tough, stubby, lateral stalk. This mushroom is not much to look at during the day, but after dark it is a showstopper. The gills and mycelium give off a greenish glow in the dark. (It might take several minutes for the luminescence to become apparent.) The species name refers to the fact that the mushroom can staunch bleeding if rubbed on a cut (similar to a styptic pencil). Older books may use the name *Panus stipticus.*

Part 2.11. Cup-like Mushrooms

These mushrooms have a small or large cup-shaped fruit body, with or without a visible stem. This group includes species of ascomycetes as well as basidiomycetes.

Part 2.11.1. Bird's Nest Fungi

Bird's nest fungi produce their spores in small packets (peridioles) that appear as "eggs" in a nest-like cup. A raindrop hitting the cup will splash the eggs several inches or even a few feet, where they will stick to leaves or attach to twigs by a long elastic cord of hyphae known as a *funiculus*. Bird's nest fungi are tiny (½ inch [1 cm] or less), and all are found on small pieces of woody debris on the forest floor and in mulch used in urban landscapes. All are widespread across North America and Europe.

Crucibulum laeve is yellow in color with a velvety exterior and is cone-shaped; its peridioles are light-colored. *Cyathus* species are dark with densely hairy exteriors and are more vase-shaped; its peridioles are dark-colored. *Cyathus striatus* features grooved or striate cup interiors; *C. stercoreus* has a smooth cup interior. *Nidularia* species (not pictured) do not produce a nest cup; rather "eggs" are released from a puffball-like fruit body.

Crucibulum laeve

Cyathus striatus

Cyathus stercoreus

Part 2.11.2. Cup Fungi

The ascomycete genus *Peziza* is probably the largest genus of brown to purple-brown cup fungi. All were once thought to be decomposers of woody debris, but there is evidence that some may be mycorrhizal symbionts with trees. *Peziza repanda* is found on woody debris throughout North America and Europe. It features a wrinkled interior with whitish fuzzy exterior when young. At maturity, this mushroom may become flattened and up to 5 inches (13 cm) in diameter. *Peziza varia* is similar and may be the same species.

There are many species of fungi that form medium- to fairly large-sized brown cups. Probably the largest and most notable is *Peziza domiciliana*. As its name implies, it can be found worldwide around human dwellings where wood, cardboard, and other materials remain wet for prolonged periods of time. On more than one occasion, I have witnessed massive fruitings with mushrooms in excess of 6 inches (15 cm) in diameter.

Gyromitra perlata (formerly known as *Discina ancilis*) is a fairly large (4-inch [10 cm] diameter) reddish-brown flattened cup, found widespread on well-rotted conifer wood. It is distinguished from pezizas by its heftier bulk, wrinkly above, veiny below, and a short but substantial stalk. Known as pig's ears, this mushroom looks quite dissimilar from other gyromitras (see page 75), but DNA sequence analysis and microscopic characteristics supports this classification concept.

Another very large cup, found in western North America and throughout Europe with hardwoods or conifers, is *Sarcosphaera coronaria*, the crown fungus. If you happen to find a young one partially buried in duff, you might think it is a hollow white rubber ball. It certainly has that look and feel. But upon maturing, this

Urnula craterium

mushroom cracks open to form a cup with a beautiful interior. And it can be quite large, exceeding 4 inches (10 cm) in diameter.

Some of the medium- to large-sized cup fungi are more goblet-shaped. Devil's urns are look-alikes for prized edible black trumpets (see page 21), but are tough and leathery, occur in early spring, and are considered inedible. Species of *Urnula* and *Neournula* are widespread. *Urnula craterium* grows from hardwood debris in the East and Europe and is about 2 inches (5 cm) in diameter and 4 inches (10 cm) tall. Also goblet-shaped and a wood decomposer, but smaller, is *Galiella rufa*, called the peanut butter cup for its chocolate and tan appearance (definitely not its taste; it is inedible). It is widespread in eastern North America.

Peziza repanda

Gyromitra perlata

Sacrosphaera coronaria

Galiella rufa

Part 2.11.3. Small Cup Fungi

There are numerous small cup fungi that you may encounter on soil or debris, and all species featured here are widespread across North America and Europe. Many are drab shades of brown, including the pezizas, but some may be quite colorful.

Caloscypha fulgens is a deeply cupped, orange fruit body typically found with coniferous debris. The outside of this mushroom is rough or scurfy, and the rimmed margin often discolors, hence the name blue-staining cup. These characteristics distinguish it from the showy orange peel fungus, *Aleuria aurantia*. Orange peel is commonly seen on bare or disturbed soil. Its cap is larger and thinner than caloscyphas, with a smoother exterior, and it is typically flattened at maturity.

Aleuria aurantia

Caloscypha fulgens

Otidea alutacea

Humaria hemisphaerica

Otidea smithii

Rabbit-ear fungi, *Otidea* species, also look similar, but upon close inspection a side slit will be seen on all caps—they do not form a complete, unbroken cup. Many are light orange to tan, including *Otidea alutacea* and *O. onotica*, but some, such as *O. smithii*, are darker.

One more notable small- to medium-sized cup to mention is the beautiful brown-haired white cup, *Humaria hemisphaerica*. This cup is easy to recognize but not often seen (it's less than an inch [2.5 cm] wide), growing on well-rotted wood or debris, often hidden among moss and other growth. The cup exterior is dark brown and hairy, contrasting with the pearly white cup surface.

Part 2.11.4. Tiny Cup Fungi

Eyelash cups (*Scutellinia setosa, S. scutellata,* and others, see page 98) are small (¾-inch [2 cm] diameter or less), bright red-orange cup fungi, usually found in very moist areas on woody debris or bare soil near wood. It is common to see them on very moist, mossy, well-rotted logs or stumps, or along forest streams. The flattened cups have dark, bristly hairs resembling eyelashes around the cup edge, but you may need a hand lens to see them. These species and other close relatives are widespread across Europe and North America and are difficult to distinguish without a microscope.

Scutellinia setosa

Chlorociboria aeruginascens

Sarcoscypha dudleyi yellow variant

Sarcoscypha dudleyi white variant

Bisporella citrina

Green stain of rotted hardwood is frequently noticed year-round. But the fruit bodies of this fungus, *Chlorociboria aeruginascens*, are much less frequently seen. If you happen upon rotted wood that has turquoise coloration running through it, examine it. The small stalked cups (¼ inch [0.5 cm] in diameter) are often on the underside of the wood, or even fissures within well-rotted pieces; green stain is widespread across Europe and North America.

Lemon drops, *Bisporella citrina*, are very tiny (usually no more than ⅛ inch [0.3 cm] in diameter), bright yellow cups found in crowded clusters on barkless hardwood branches and logs. Although minute, this fungus is frequently noticed due to its bright color and gregarious nature; it is widespread across Europe and North America.

Sarcoscypha species are very commonly seen, small (1–2 inch [2.5–5 cm] diameter) bright red cups that show up in the late fall through spring on partially buried wood. *Sarcoscypha dudleyi* is eastern and *S. coccinea* is western, and there are others. Both are considered European. Most have very short stalks, but some have well developed stalks that may be buried. And although they are normally red, occasionally you will find white or yellow variants.

Part 2.12. Ball-shaped Mushrooms

These fruit bodies are roundish and can range from very small to enormous. Very small ball-shaped fruit bodies can resemble small plant fruits, plant galls, insects, or slime mold fruit bodies.

Part 2.12.1. Large Puffballs

Giant puffballs are one of the "foolproof four" safe edibles and are collected by many. All true puffballs are safely collected if they are pure white throughout. If your puffball is any other color inside, or if it is over-mature and squishy, it's not a safe edible. When mature, the puffball skin (perideum) tears or breaks open to reveal the brown spore mass inside. Large puffballs are decent, if bland, edibles and mostly take on whatever flavors they are cooked with. Small puffballs (see page 101) are also edible but not really worth the trouble.

The surface of the giant puffball, *Calvatia gigantea* (also called *Langermannia gigantea*), is smooth and white when fresh, becoming olive to brown when mature. This fruit body is very round. When it is mature, the surface splits open to release the spores. Average specimens are about the size of a football (12 inches [30 cm] diameter), but they can be basketball-sized or larger. This species is found in open woods or in fields or pastures in late summer and fall, and it may even fruit in fairy rings. It is widespread and common across eastern North America and Europe.

The giant puffball of western North America is *Calvatia booniana*, noted by its more elongated shape and a perideum that cracks and peels. The beautiful purple-spored puffball, *Calvatia cyathiformis* (see page 100), can be quite small or nearly as large as the giant puffball

Calvatia gigantea

Calvatia cyathiformis

Calvatia cyathiformis dark at maturity with backpack for size reference

Calvatia brown spores at maturity

at maturity. The base of this mushroom has a prominent lower portion that tapers gradually to a somewhat pointed base. It will become dark brown at maturity and, as its name indicates, it has a spore mass that is purplish at maturity. *Calvatia craniiformis* is very similar, but the interior never turns purple. Both are found across North America and reportedly southern Europe.

Earthballs and earthstars are not puffballs. They are not white inside, and many are inedible or toxic and should be avoided. Furthermore, stinkhorns and deadly amanitas start out in an "egg" stage; be sure to refer to those pages. If you slice a puffball in half and find a mushroom inside, it's not actually a puffball (it is likely an immature stinkhorn or *Amanita* mushroom; the latter could be deadly).

Part 2.12.2. Small Puffballs

As mentioned previously, all true puffballs are safe edibles. Small puffballs are commonly encountered, but they are not really worth the effort of collecting because they are small (usually 2 inches [5 cm] in diameter or less) and not particularly palatable. Probably most common is the pear-shaped puffball, *Lycoperdon pyriforme*. This puffball is hard to misidentify as it is usually found on decaying wood. When immature, it has very prominent white mycelial strands attached to its base. As it matures, it enlarges and develops its characteristic inverted pear-shaped appearance. As with other *Lycoperdon* species, the spores are "puffed out" through a pore in the top. (Puffball squeezing can be irresistible to children!) Curiously, the genus name *Lycoperdon* translates to "wolf fart."

Lycoperdon pyriforme

Lycoperdon echinatum

Lycoperdon nigrescens

Lycoperdon perlatum

The gem-studded puffball, *Lycoperdon per-latum*, is usually found on the ground. Its name comes from the numerous spines on its surface. This fruit body is white to cream and slightly larger than the pear-shaped puffball, although it has a similar shape. Both are common across Europe and North America.

Seen in large troops, small puffballs can be quite photogenic. The showiest species, argu-ably, are *Lycoperdon echinatum*, the spiny puff-ball, and *L. nigrescens*, the dusky puffball. The spiny puffball is found on the ground on decay-ing litter in the forest and in landscaping mulch in eastern North America and across Europe but is not especially common. The small fruit body (1 inch [2.5 cm] diameter) is white when immature and then gradually darkens to buff or light brown as it matures. The spines are actu-ally pointed tufts of hairs. The dusky puffball also occurs on soil in western North America and throughout Europe. It is easy to distinguish by its very dark gray-brown to chestnut-brown coloration and very short spines.

Part 2.12.3. Puffball-like Mushrooms: Earthstars, Earthballs, and False Puffballs

Earthstars greatly resemble puffballs. But notice that their outer skin (peridium) splits in a very distinctive star-shaped pattern, expos-ing the inner spore sac—essentially a small puffball. The points of the "star" often re-curve toward the ground and can actually push the inner puffball off the ground, as with *Geastrum fornicatum*. This feature gives earthstars a very animal-like appearance, especially if the pro-cess is viewed using time-lapse photography. Most common across Europe and North America are species of *Geastrum* and *Astraeus*, all of which occur on woody debris and are about 1–3 inches (2.5–8 cm) in diameter.

Sclerodermas go by many names, including pigskin puffball, poison puffball, and earthball. The most common species across North America and Europe is *Scleroderma citrinum*. This globose- to egg-shaped false puffball

Geastrum fornicatum

Scleroderma citrinum

(about 3 inches [8 cm] in diameter) may be found on decaying wood or in mossy areas, but it can also be found on bare ground. Sclerodermas are mycorrhizal with hardwood and conifer trees nearby. When immature, the interior might be white, but it gradually darkens to purple-black. The thick peridium is rough (like a football) and is also fairly tough. When the earthball is mature, the skin splits open and the spores are released. Do not confuse the pigskin puffball with edible puffballs, as it is poisonous!

Geastrum sessile

Part 2.13. Coral-like Mushrooms

Coral fungi look like undersea creatures. They usually grow saprobically on decaying woody debris and logs, but many arise from soil. While some are drab colored, many are brightly colored (red, yellow, orange, and purple). Species can be highly branching, barely branching, or entirely unbranched as worm corals are. Their size is variable as well, ranging from a few inches tall and wide to massive clusters more than 12 inches (30 cm) in diameter. Spores are borne directly from the side of branching stalks.

Part 2.13.1. Branched Coral Fungi

Coral fungi are numerous and commonly encountered, but they are mostly difficult to identify as species. Unless noted, all the species shown here are found widespread across North America and Europe.

Probably the most widely seen coral fungi is the crown-tipped coral, *Artomyces pyxidatus.*

This coral (also known as *Clavicorona pyxidata*) is whitish-buff when fresh, and it gradually ages to a dull, buff-yellow. It occurs on logs and fallen limbs in deciduous woods. Note its highly branching growth, like a candelabra, and crown-like tips, which help to differentiate it from common *Ramaria* species, such as this green-staining *Ramaria abietina* from western North America. Ramarias may occur on rotting wood, but some are mycorrhizal partners with living tree roots, and they come in just about any color.

If you find a striking purple coral, it's likely the violet coral, *Clavaria zollingeri.* This beauty may be found in woods, the margins of woods, and even in grassland areas.

False coral fungi (*Tremellodendron* species) could be confused with true corals but are very tough and rubbery, and they typically occur in woodland areas on very compacted or disturbed soil.

Artomyces pyxidatus

Ramaria abietina

Clavaria zollingeri

Part 2.13.2. Unbranched Worm Corals

Worm corals are more demure than their cousins on the previous page. Although some may occur in gregarious patches, a single fruit body will be a singleton or small cluster; the tallest species will be 3–5 inches (8–13 cm) tall. All are considered saprobes of forest debris. They are fragile, breaking if handled. All the species shown here are found widespread across North America and Europe.

White worm corals, *Clavaria vermicularis* and *C. fragilis*, are commonly seen in woodland habitats. They are bright white but yellow at the tips as they age. *Alloclavaria purpurea*, purple club coral, is also quite common, showing up as discreet clusters, but more often than not in very large troops. The color can be variable as well, ranging from violet to a dull smoky gray-violet.

Yellow worm corals, *Clavulinopsis fusiformis* and *C. laeticolor*, are similar in appearance. Both are yellow to yellow-orange in color, but note that the tips of *Clavulinopsis fusiformis* become pointed and very dark with maturity. *Clavulinopsis laeticolor* does not grow in clusters (they may be very loose clusters), unlike the clustering *C. fusiformis*.

Clavulinopsis fusiformis

Alloclavaria purpurea

Clavaria vermicularis

Part 2.14. Simple Club-shaped Stalk Mushrooms that Arise from Soil or Wood

At first glance, club fungi look somewhat like coral fungi. Given their drab colors, you may overlook them altogether. Club fungi mostly appear to be growing from soil, but they are more likely to be growing from a living (or recently dead) host plant, an arthropod, or even a truffle.

Part 2.14.1. Club Fungi on Wood

Xylarias look like something from another world. These fungi are ascomycetes that produce their spores in chambers (perithecia) lining the surface of the stromal "fingers." Although their fruit bodies seem to be coming from the ground, they are getting nutrients from buried wood. At maturity, the fruiting bodies appear much like coral fungi (see pages 104–105), but are hard and dark. During the spring and when these fruit bodies are immature, they may appear white near the tips. This is because the stroma begin their cycle by producing lighter colored asexual spores (conidia). The dark sexual spores are released from the perithecia later in the season.

Dead man's fingers, *Xylaria polymorpha* and *X. longipes*, are the largest members of the group. An individual stroma may be 4 inches (10 cm) above soil (but in the case of *X. polymorpha*, it may be connected to a larger base deep in the soil; see photo). The two species are similar, but *Xylaria longipes* typically has a very thin stalk base and often grows singly. Both are com-

Xylaria polymorpha

Xylaria polymorpha immature / white

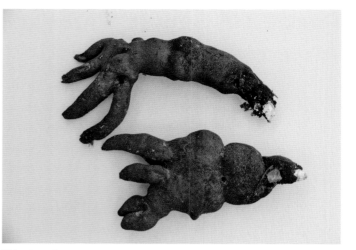

Xylaria polymorpha excavated

Xylaria hypoxylon

mon in eastern North America and Europe.

Carbon antlers, *Xylaria hypoxylon*, are found across all of North America and Europe. Also known as candlesnuff fungus, this species is distinguished from the previous two by its very thin stroma. The pointed stromata are often forked at their tips, giving them the appearance of antlers.

Part 2.14.2. Club Fungi on Soil

The club fungi on this page all require a living host. Some, such as the sweet club, *Clavariadelphus pistillaris*, are mycorrhizal with trees, and they are attached to tree roots deep in the soil. This basidiomycete fungus is the largest species of the group (6 inches [15 cm] or more) and is quite widespread throughout eastern North America and Europe; in western North America, *Clavariadelphus occidentalis* is similarly large. *Clavariadelphus truncatus* ranges across North America and Europe and is distinguished by having a flattened top.

The *Cordyceps* group of fungi share the general look of the sweet clubs but are distantly related ascomycetes that parasitize other plants and animals. The largest and most commonly seen species in North America and Europe seem to be growing from soil, but upon careful excavation, the collector may be delighted to find them attached to a deer truffle (*Elaphomyces* species). The fruit bodies function similarly to xylarias (see page 107), in that they produce their spores in chambers (perithecia) that are embedded in the fertile area of the club.

Elaphocordyceps (Tolypocladium) ophioglossoides

Clavariadelphus pistillaris

Elaphocordyceps (*Tolypocladium*) *capitata* **close-up**

Cordyceps **on spider**

Elaphocordyceps (*Tolypocladium*) *capitata* has a discreet head atop the stalk where spores are produced; the fruit body is directly attached to a deer truffle. In contrast is *Elaphocordyceps* (*Tolypocladium*) *ophioglossoides*, which has an elongated head that is similar in appearance to the rest of the stalk; it is attached to a deer truffle by yellow hyphal cords (and they may run several inches deep in the soil).

It takes a keen eye to spot *Cordyceps* species that parasitize and kill insects. In some cases, the victim may be perched on a plant leaf out in the open, such as the spider pictured. Other times the insect may be in the puparia stage and buried underground, in which case the fungal fruit body may resemble small matchsticks poking from the soil.

Elaphocordyceps capitata

Part 2.15. Jelly-like Mushrooms

There are many kinds of fungi that grow from wood and produce fruit bodies that look like jelly drops or convoluted, brain-like masses of jelly. Some are ascomycetes and others are basidiomycetes.

Part 2.15.1. Jelly Fungi

All species discussed here are commonly seen across North America and Europe. Many are small and come in bright colors such as orange or yellow. *Dacrymyces chrysospermus* is a yellow-orange jelly found on conifer wood. It resembles the bright yellow witches' butter, *Tremella mesenterica*, which is found on hardwood (often associated with *Stereum ostrea*, the false turkey tail, page 85). *Exidia recisa* is a black jelly and *E. alba* (also known by the tongue-twisting name *Ductifera pululahua*) is a white jelly. *Ascocoryne sarcoides*, purple jelly drops, are frequently seen on conifer and hardwood, often on the cut end of logs. The fruit bodies are small, ½ inch (1 cm) or less. When young they appear as violet jelly drops, but they mature to resemble flattened cups that may crowd together.

Some jellies are larger and more developed, sometimes resembling ears or clusters of ears. *Auricularia auricula*, the wood ear or tree ear,

Dacrymyces chrysospermus

Leotia lubrica

Ascocoryne sarcoides

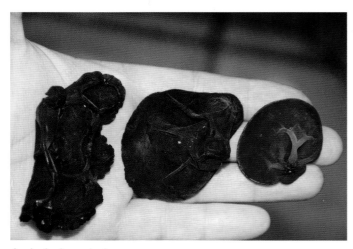

Auricularia auricula

is found on rotting wood throughout the year, though primarily in the spring and fall. The appearance of some specimens is virtually identical to an ear (*aurical* is Latin for "ear") and can grow an inch (2.5 cm) to several inches wide. *Auricularia* is cultivated in Asia and is frequently used in Chinese and Japanese cooking. This species can sometimes be mistaken for pig's ear (*Peziza* species), which are ascomycetes, or *Tremella foliacea*.

Jelly babies (*Leotia* species) are fairly large (up to about 3 inches [8 cm] tall) mushrooms featuring a slender stalk with a club-shaped cap. They have a nice jelly feel about them—that is, if you can pick them; they are very slippery. *Leotia lubrica* is yellowish to light tan and *L. viscosa* is similar but with a dark green cap.

Part 2.15.2. Jelly-like Fungi

The fungi on this page share characteristics with jelly fungi and coral fungi, and all are commonly seen across North America and Europe. Yellow tuning forks, *Calocera cornea*, look like many small jelly-like horns erupting from fallen hardwood (and occasionally conifer) logs.

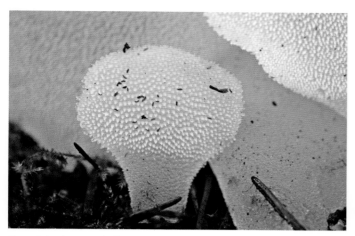

Pseudohydnum gelatinosum

Calocera cornea

Although small (less than 1 inch [2.5 cm]), they are hard to miss, given their bright colors and large troops. These little fruit bodies are typically unbranched but may be forked; eruptions may be small clusters or singletons. Looking more like a coral fungus, *Calocera viscosa* is more branched and slightly larger (1 inch [2.5 cm] tall, with clusters about 4 inches [10 cm] wide) than its cousin. Known as false coral, it is commonly seen on buried conifer (occasionally hardwood) debris.

Pseudohydnum gelatinosum, cat's tongue, also grows from well-rotted wood in moist areas and has a jelly-like feel. But cat's tongue is somewhat translucent—white when very young and becoming more tan color with age. If you look closely, you will see very small teeth hanging down from the cute little mushroom.

Calocera viscosa

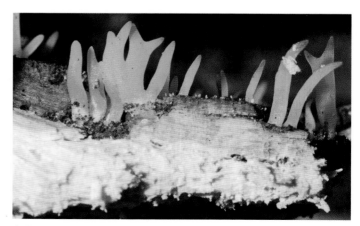

Calocera cornea close-up

Part 2.16. Crusts and Mushroom Oddities

This final section is a grab bag of strange fungi. Although strictly speaking you would not think of them as mushrooms, they are certainly fruit bodies of macrofungi, and they are frequently encountered in the woods and sometimes around the home or garden. One is even a choice edible featured in the Culinary section.

Part 2.16.1 Crust-like Fungi

There are numerous parchment or crust-like fungi that occur on wood, and most are pretty hard to distinguish. As a group, they are often overlooked by those interested in mushrooms. And this is too bad, as many play interesting roles in the environment as decayers and some are symbionts with living trees. Some may be destructive of urban dwellings. Most crusts grow over large areas of fallen logs, often on the moist underside, close to soil. It would seem a difficult place from which to disperse spores into the air column, but it was recently shown that some entice very small insects and other arthropods to do the work for them.

Velvet blue spread looks like striking patches of blue velvet on the surfaces of deciduous sticks and logs on the forest floor. The basidia and basidiospores are borne directly on the velvety surface. This beautiful fungus is hard to mistake for anything else and occurs across North America and Europe. It goes by the scientific names *Pulcherricium caeruleum*, *Terana caerulea*, and *Corticium caeruleum*.

Ceramic parchment, *Xylobolus frustulatus*, is a perennial (present all year) wood decay fungus that has the appearance of cracked and caked mud on logs and stumps, particularly on the cut surface. This fungus is quite common across North America and Europe, and is closely related to *Stereum* species, which also occur on logs.

Terana caerulea

Crowded parchment, *Stereum complicatum*, can cover entire dead branches and logs, and is very similar to the false turkey tail (*Stereum ostrea*, page 86). Crowded parchment is smaller, has a wavy or ridged cap, and has brighter orange-tan zones.

Trembling merulius, *Phlebia tremellosa*, and the similar looking *P. incarnata* are found on well-decayed wood, and they have pinkish-orange, somewhat gelatinous, spore-bearing surfaces. Phlebias have ridges in a netlike pattern that superficially look like pores of a polypore. The edges of the fruiting body are often reflexed into a shelf-like growth with a whitish, somewhat hairy upper surface. This curious mushroom has gone by the name *Merulius tremellosus* because of its merulioid ("*Merulius*-like") wrinkled surface. Dry rot fungus, *Serpula lacrymans*, is very similar looking but a bit more olive-brown to rusty-brown. It occurs on decaying conifer wood in nature but is rarely seen. It is much more commonly seen (and feared) inside human dwellings, where it can penetrate brickwork and decay dry wood after getting a start where water damage has occurred.

Xylobolus frustulatus

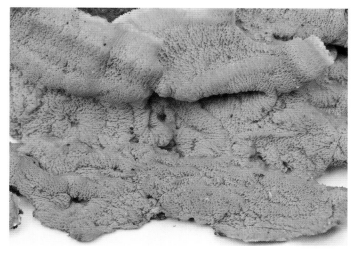

Phlebia tremellosa

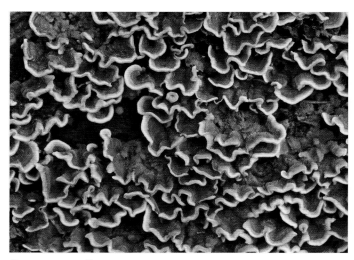

Stereum complicatum

Part 2.16.2. Rusts and Smuts

Rust and smuts are basidiomycetes and thus close relatives of mushrooms. All are parasites of plants and mostly pretty small. The rusts and smuts are large and fascinating groups of fungi. The rusts are especially interesting, as many require two different plant hosts to complete their life cycle. Many rusts are pests of agricultural crops.

Cedar apple rust, *Gymnosporangium juniperi-virginianae*, is a plant pathogen that results in weird alien-like lifeforms on plants. This fungus, and other close relatives, are widespread across North American and Europe. They may appear as jelly-like projections from the stem or branches of living tree host plants, or as ball-like galls with brightly colored jelly projections. Cedar apple rust may be seen wherever apples or crabapples (*Malus*) and junipers coexist across Europe. In eastern North America, the fungus is common on eastern red cedar (*Juniperus virginiana*) and can be a destructive or disfiguring disease on both the apples and cedars. Quince and hawthorn are also hosts. This fungus may look a bit like a jelly or *Calocera cornea* (see page 112), but those all occur on dead wood.

Corn smut, *Ustilago maydis*, is conspicuous and important historically. It has long been a popular edible fungus in Mexico and is now gaining fame in mycophilic circles as haute cuisine. Corn smut is a parasite on living corn (maize) plants and may be seen throughout North America and Europe, especially in the warmer regions. Historically, this fungus was common on field and sweet corn, but modern varieties of corn have been bred for resistance to it (as is the case with field corn). Heirloom corn is still susceptible, as is popcorn and Indian corn. It's worth noting that just about any corn can become infected if injury occurs (from hail, for example) during mid season. The fungus infection results in galls that develop on ears, stalks, and tassels of the plant. The galls are edible and have long been considered a delicacy in Mexico, where they are known as huitlacoche (or cuitlacoche) and Mexican corn truffle. An older name is *Ustilago zeae*.

Ergot, *Claviceps purpurea*, produces hard, black bodies known as sclerotia on rye or other cereal grains and grasses. These fruit bodies are small and rarely noticed. On occasion this fungus is collected at wild mushroom forays where experts are on the lookout for it (see photo). Ergots are widespread across North American and Europe. This fungus is quite poisonous. One of the compounds in ergot is closely related to LSD and is responsible for hallucinations. Consumption of ergot sclerotia on contaminated grain and bread flour in the Middle Ages resulted in localized outbreaks of hallucinations, convulsions, and gangrene. These episodes caused peasants to believe they had been visited by something known as Holy Fire or St. Anthony's Fire. Some historians believe that the Salem witchcraft incidents and the French revolution resulted from ergotism (ergot poisoning). Another ergot compound causes constriction of blood vessels and is now processed into various drugs (such as Cafergot) for the treatment of vascular headaches.

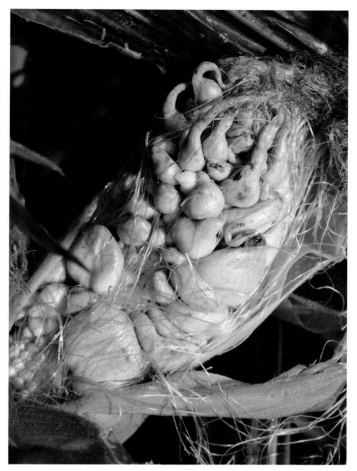

Ustilago maydis

Claviceps purpurea

Gymnosporangium juniperi-virginianae

PART III
Mushroom Cultivation

Some mushrooms are mycorrhizal partners of trees and other plants, as we have discussed. Mycorrhizal fungi cannot be cultivated. Many other mushrooms, however, are saprobic in nature—they decompose organic material—and those can be cultivated.

Mushroom cultivation relies on the basic principle of introducing an organism to a food source. Mushrooms and other fungi digest organic material and draw that nutrition to a reproductive structure that we call a mushroom, or a fruiting body. It is the reproductive part of a fungus. The fungal mycelium secretes enzymes to break down the nutritious matter it encounters.

To cultivate mushrooms at home, you can start with a mushroom, mushroom spores, or spawn (a culture of the fungus, typically on grain or sawdust). Spawn can be purchased, and it is a great option for beginners. For the advanced mycophile, there are multiple ways to make your own spawn at home, starting with a piece of mushroom or spores.

Mushroom Cultivation on Household Compost

You likely have a compost pile at your home where grass clippings, fall leaves, and garden scraps go. Why not employ mushrooms to help break it down? *Agaricus* species will work if the matter is well-composted already. If it is not broken down, try wine caps or blewits. You can start with pieces from freshly collected mushrooms (the trimmed ends of stalks) or from spawn. Spawn can be purchased and works very well.

To prepare the site, just rake back a layer of compost. Add your spawn and cover with compost. Water the soil to help keep moisture levels up and to help speed up the spawn run. The mushrooms will fruit in late fall or even winter in warmer climates. If one type of mushroom fails to work, don't give up; just try another species to find the one that works best on your type of compost.

Growing Blewits on Composted Leaves

Blewits grow naturally on leaf litter on the forest floor. They are typically a fall mushroom, fruiting very late in the year. Starting a bed of blewits in leaf litter is an easy way to get great, edible mushrooms after much of the mushroom season is over. They can appear by the hundreds at times and are a favorite of experienced mushroom hunters. If you have a pile of leaves from last year at your disposal, you have a great substrate for blewit cultivation.

For fall inoculation, you can start with pieces from freshly collected mushrooms (the trimmed ends of stalks). Just place them in your leaf pile. Other times of the year, spawn can be purchased and works very well. For summer inoculation, find an area in the shade to start the bed.

Rake away debris to expose the soil surface. A heavy watering at this point will give you a nice source of humidity as the mycelium develops. Using leaf litter from last fall, cover the prepared area to a thickness of about 2 inches (5 cm). Blewits are very aggressive, and sterilizing the substrate is not necessary. Any leaves will work, but if you are going to use pine needles, mix them with the leaves of other trees. Add blewit spawn to the leaf litter. Stir it around, breaking up all the chunks. Cover the spawn layer with another layer of leaves. Give the bed a good watering. Don't oversoak it and drown the mycelium, but give it enough water for the leaves to stick together.

You can cover the bed with a sheet of clear or white plastic for a couple of months to help get the colony going, but it is not absolutely

necessary. Black plastic can cause the bed to overheat and potentially kill the mycelium, so avoid it. Peel the plastic back and water about once per week if covered. Blewits appear in large troops in the fall and will grow very quickly.

Inoculation of Straw

There are two basic methods of inoculating straw. The first is used when the straw is isolated in its own environment, and the second involves direct soil contact. Oyster mushrooms and blewits do not require soil contact, so they grow well in containers or plastic sleeves. Other mushrooms, such as wine caps, require contact with soil to grow. These mushrooms are typically grown outdoors or in a greenhouse.

Growing Oysters and Blewits on Straw in a Container

Using straw for a substrate on which to grow mushrooms can be fast and effective. Colonization is far quicker with straw than it is with other substrates, and it may produce mushrooms as soon as 2 weeks after inoculation. It is a cost-effective way to obtain high yields from little material. The mushrooms will be cleaner than they would be if they were grown on wood, and they require little preparation after harvest.

Many different types of containers can work for the cultivation of oyster mushrooms on straw. The purpose of the container is to hold the straw together while offering openings for oxygen to use to reach the mycelium. Plastic bags work well for this, but any container can be

made to grow oyster mushrooms. A plastic pail can be drilled to allow oxygen to reach the sides of the straw pack and for mushrooms to begin fruiting. A flowerpot can be pierced in the same fashion, becoming a decorative oyster-growing container to hang in or near a garden. Bales of straw can be found at most garden or farm supply stores.

To prepare the straw for the cultivation of mushrooms, it can be a good idea to run the straw through a leaf shredder to increase the surface area and to make the straw easier to handle. This is not a necessary step, but it does make packing the containers an easier task. It also gives the mycelium an advantage by putting more straw in contact with the spawn. In lieu of a leaf shredder, home growers also use weed whackers or push mowers to shred up straw nicely. Plan ahead: Do this in a place where you intend to cultivate from straw in contact with soil (see below), and it will make for less clean up.

Whole or chopped straw must be pasteurized to eliminate contaminating microbes. Most hobbyists will have great success with cold pasteurization. Soak the straw in water for 3 to 5 days. Drain the straw for 24 hours to prevent excessive water in the containers.

Once the straw is drained, you are ready to inoculate.

Wash your hands thoroughly before handling the spawn and straw substrate. Spores or bacteria that may be present on your hands can colonize the straw and outrun the oyster mushroom mycelium. You can use trimmed pieces of oyster mushrooms or purchased spawn. Trimmed pieces of store-bought oyster mushrooms can work if they are very fresh. Purchased spawn will likely work much better as you can use lots of it (the higher the inoculation rate, the faster the mycelium will colonize the substrate, and the faster you will see mushroom production), plus it comes sterile from the factory. If you are using spawn, sprinkle it in an even layer over the top of straw. After each application of spawn, roll and fold the straw. Fold the straw by sliding your hands underneath it, palms up, after an application of the spawn. Lift your hands up and turn your palms down, folding the straw over on itself. When the spawn has all been mixed in evenly, the prepared substrate is ready for containers.

Sterile plastic sleeves are the recommended containers for oyster production, but plastic buckets or flowerpots with holes drilled into them are popular too—anything can be made to work. Make sure whatever you use gets washed and dried first. Pierce holes in the pot to allow an opening for the mushrooms to fruit.

Start with a thin layer of prepared substrate on the bottom of the container. Pack it in, making sure to fill any corners of the container. The idea is to eliminate all air from the inside of the vessel to promote fruiting at the locations you created. With the plastic sleeves, pack both corners first as tightly as possible. It can be difficult to get a good solid mass at the bottom of the bags. Adding small amounts to start with will make this easier. Fill the container as much as possible.

For a solid container, you will have to add a lid. If you don't have a lid, a layer of aluminum foil and a rubber band as a clamp can work well. Use cable ties for the plastic sleeves and tighten them as much as possible. Use a knife to slice fruiting holes on the plastic sleeve. Make an opening about every 6 inches (15 cm) around all sides of the sleeve. This is the escape hole for the mushrooms to fruit.

The next few weeks will be the incubation of the oyster mushroom mycelium. They require a humid environment and at least some light. For

the more colorful strains of oyster, more light should be allowed. This will improve the color of the mature mushroom. Humidity and temperature affect incubation times. A room that is 85% humidity is ideal for oyster production.

The colonization will take anywhere from 3 to 10 weeks. You will start to see the straw change color in just a couple weeks. This is the mycelium colonizing the straw. Young mushrooms will start to form at the fruiting vents soon after complete colonization. These "pins" will develop quickly into mature mushrooms.

Oyster pins will only take a day or two to grow to maturity. They develop very quickly and should be harvested immediately. If left on the container, they will begin to rot rapidly. This can put the container at risk for contamination and affect future fruitings.

Oysters are easy to harvest. Simply grab the mushroom by the base and twist it off. There will be straw debris connected to the base; it can be trimmed with a knife or picked off with your fingers. The entire oyster mushroom is tender and edible, so don't trim off more than necessary.

Growing Wine Caps on Straw on the Ground

Wine caps (*Stropharia rugosoannulata*) are a great mushroom to grow on the ground with little chance for failure. They are a large, flavorful mushroom that will tolerate the high temperatures of summer. The process is relatively simple and can produce mushrooms quickly.

Place a wine cap bed in an area that is shaded for at least most of the day. Prolonged periods of direct sunlight will overheat or dry the bed and damage the fungus, possibly killing it. Water the ground thoroughly before you start. Remove any green matter, so all the debris in the bed is dead. A great idea is to simply rake a compost pile to the side, exposing the bare ground beneath it, and start your bed there. Leaving twigs and dead leaves is acceptable, as the wine cap mycelium will colonize that as well and use it for additional nutrition.

Never substitute hay for straw. Hay composts far too quickly, and you will not get any mushroom production.

Because you are not going to be packing it into small containers, it is not necessary to shred the straw, as above. Start by soaking it for 3 to 5 days. This can be done in a large container: Fill the container with water so that all the straw is covered. Then add weight to the straw to ensure all of it is submerged. Remove the straw from its container and let the excess water drain off for a few minutes. It is not necessary to drain the straw completely.

Shake the straw onto the bed so that it is rather fluffy when it piles up. The first layer should end up being about 4 inches (10 cm) thick and as even as possible. Break apart any large clumps to ensure the most exposed surface area as possible. The next step is to add an even layer of spawn to the first layer of straw.

Grain or sawdust spawn can be purchased. Spread it as evenly as possible and break up any clumps that may fall into the straw. The finer the sawdust is sprinkled, the more contact between spawn and substrate will occur. This will ensure a fast spawn run and thorough colonization.

The bed is now ready for the second layer of straw. Remove another batch or straw from the water and let drain for a few minutes. Shake the straw as you did in the first layer and spread it evenly over the entire bed to a depth of about 2 inches (5 cm). Remember to break up any large clumps of straw to keep the substrate as even as possible. This will ensure a cosmetically appealing bed and keep spawn run efficient through the entire bed. Use the back of a lawn rake to tamp the layers together. This will increase contact between the spawn and the straw and provide a better rate of colonization. Any spawn that does not contact straw will eventually die, and this will slow down the colonization process. You may choose to repeat with another layer of spawn and a final layer of straw on top. When inoculation is complete, give the bed a good watering.

After the last dressing of water is applied, the bed is ready for its cover. Place a clear plastic sheet cut to size over the entire bed and weigh it down around the edges. This will prevent evaporation of the valuable water that we applied in the last step. Do not use black plastic, as it may overheat and kill the fungus within the bubble. This plastic cap should be left in place for 4 weeks.

Inspect the bed about once per week to ensure it is staying wet. If the bed seems to be drying out, add 2 gallons (7.5 L) of water. Adding too much may drown the mycelium. You should see droplets of water on the inside of the plastic bubble.

At the end of the 4-week incubation period, remove the plastic sheet and water the bed. Peel back the straw in different places around the bed to check for spawn run. This will be a white, stringy mat with a very sweet smell. Replace the straw as close as you can to where it was. If there is no spawn run, or if it appears to be weak, replace the plastic for another week or two and continuing the watering regimen. The mushroom should start fruiting 3 to 5 weeks after the plastic sheet has been removed. The first mushrooms you will see will be around the outer edge of the bed, but there are others hiding in the deeper straw. They can be difficult to find. Look for irregular bulges in the straw.

PART IV
Culinary Uses and Preservation

This part is all about what to do with your favorite wild and cultivated mushrooms, now that you have them. We'll discuss how to prepare and preserve them, after which we've included some of our favorite mushroom recipes.

Processing and Preservation

Most wild and cultivated mushrooms can be kept for about 1 week in the refrigerator. They should be stored in paper, never in plastic. Or, simply place a clean cotton towel in the bottom of the crisper drawer, add your mushrooms, and cover them with another cotton towel. Dry towels will wick away any moisture and prevent molding. Never place wet mushrooms in the fridge; air-dry them on a cotton towel first.

If you have more mushrooms than you can use up in a few days, consider preserving them for later use. Drying is an excellent way to preserve (and display) them, and some species are as good as or better dried than they are fresh. Freezing is also suitable for many kinds of mushrooms. Both techniques are simple and explained below.

Not all mushrooms preserve well, however, and these really should be used up while fresh. Matsutake mushrooms lose their wonderful aroma if dried; freezing preserves this somewhat, but the crisp texture is mostly lost. Chicken of the woods, or sulfur shelf, loses all texture, aroma, and color when preserved by any means. Shaggy manes and inky caps lose everything and turn to a bowl of ink if not used up quickly. They contain powerful enzymes that cause the caps to pretty much digest themselves. If you must store any inky cap mushrooms overnight, try this trick: Store them submerged in a bowl of water in the fridge. They may last for a few days if all air is excluded from them, which suppresses the enzyme activity. And don't worry about them becoming waterlogged before cooking, as inkies are mostly water anyway.

HOW TO PRESERVE MUSHROOMS

Here is a handy guide to the best ways to preserve your wild and cultivated mushrooms. Recommended (++), mediocre (+), and not recommended (blank) are the opinions of the authors.

Mushroom	Fresh	Dried	Frozen	Pickled
Agaricus spp., Field Mushrooms	++		+	+
Black Trumpets	++	++		+
Blewits	++		+	+
Boletes	++	++	+	+
Candy Caps		++		
Chanterelles	++		+	+
Chanterelles, Yellow-foot	+	+	+	+
Chicken of the Woods, Sulfur Shelf	++			

Mushroom	Fresh	Dried	Frozen	Pickled
Hedgehog Mushrooms	++		+	+
Hen of the Woods	++	++	+	++
Honey Mushrooms	+			+
Huitlacoche, Corn Smut	++		++	
Lobster Mushrooms	++		+	
Matsutake, Pine Mushrooms	++			
Milk Mushrooms	++	+	+	++
Morels	++	++	+	
Oyster Mushrooms	++		+	+
Puffballs	+			
Shaggy Manes, Ink Caps	+			
Shiitake	++	++	++	+
Tooth Fungi, Lion's Mane, Hawk's Wing	++			
Wine Caps	++		+	

Cleaning Mushrooms

Mushrooms should never be soaked. If they are far gone or absolutely filthy, they should be left in the woods. Most mushrooms will be damaged or ruined in water. Some very sturdy ones (e.g., chanterelles and young morels) can survive washing, but most others will be damaged or destroyed.

Clean mushrooms begin in the forest. When picking them, take extra time to remove debris from edibles, trim off dirty parts of stems and bad spots, and place the mushrooms in a clean basket. And keep it clean: Do not pile dirty mushrooms on top of clean ones. This is especially important for gilled mushrooms! Dirt and grit that has gotten into the gills of mushrooms are impossible to clean out. I personally have zero tolerance for crunching grit in my food. Take two baskets if you need to.

It should go without saying that mushrooms riddled with holes from bugs should be tossed out. A soft brush in the kitchen or on your mushroom knife is invaluable; a damp cloth in the kitchen can wipe away any last traces of dust or dirt. If you absolutely must wash a dirty mushroom, air-dry it on a towel for a few hours before storing or cooking it. Wet mushrooms mold in the fridge and do not sauté in the skillet; they boil.

Freezing Mushrooms

Freezing is a great way to store many kinds of mushrooms; just follow the steps below to ensure the final product is the highest quality possible. All mushrooms to be preserved should be fresh and firm.

A good way to ensure your frozen mushrooms will maintain good flavor and texture is to cook them before freezing them. They do not need to be cooked thoroughly, but they should be cooked at least partially. Cooking deactivates enzymes and preserves texture and flavor in mushrooms; it's just as vegetables such as corn or beans must be blanched before they are frozen to prevent them from degrading in quality. The freezing process is the same for all mushroom species: Simply slice them and then either blanch them in boiling water for a minute or lightly sauté them. I often do this with mushrooms as soon as I bring them home, saving the sautéed mushrooms in a big bowl until dinnertime. All I need to do next is throw a large handful or two into the dish I'm preparing, and then the rest can go into a small zip-top bag and straight into the freezer.

To freeze large batches of mushrooms, slice and arrange them separately (not touching) on a baking sheet lined with waxed or parchment paper. Stack the baking sheets in the freezer overnight. Once the mushrooms are frozen, transfer them to freezer bags. Freezing them separately like this will allow them to hold their shape better and prevent from becoming a single frozen mass. When you're ready to cook, remove a portion and return the bag to the freezer. Allow frozen mushrooms to thaw and warm them back up or add them to a dish that is already cooking. Frozen mushrooms retain much of their flavor and texture and can be used just as fresh mushrooms would be.

Drying Mushrooms

Drying is the preferred method of storage for most mushrooms, but it works better for some than it does for others. Morels, candy caps, and shiitake mushrooms dry extremely well and can be kept for years. Oyster mushrooms and wine caps dry readily, but the texture of the reconstituted product can be a little soft for most people.

Chanterelles dry nicely but rehydrate to be tough and leathery. In cases like this, where the dried mushroom's texture may not reconstitute well, consider grinding the dried mushrooms into a powder. Mushroom powders often retain

Dried morels

Dried morels

eat the toxic mushrooms. It's best not to mix toxic and edible species on the dryer at the same time.)

Mushrooms can also be air-dried with a fan, hung up to dry with needle and thread, or sun-dried on screens outside. None of these methods work as quickly or as well as a dehydrator. Most places in North America and Europe have a good deal of humidity in the air, so while your air-dried mushrooms seem dry, they are not likely dried enough. I cannot tell you how many times I've had friends pull out jars of air-dried mushrooms only to find that they'd grown moldy. Also, air-dried mushrooms shrink and do not look as nice in a prepared dish as dehydrator-dried mushrooms.

The basic principle of drying is that air flow is more important than heat. All mushrooms should be dried at a temperature of 110° F (43° C) *or less*. On your dehydrator, this will be the lowest setting. Do not purchase it if the lowest setting exceeds this. This will ensure that the mushrooms dry, rather than cook. Cooking the water out of the mushroom will dry it, but the finished product will lose its fresh flavor and pleasant color. Mushrooms dried with too much heat will darken.

A well-dried mushroom will retain most of the characteristics of its fresh counterpart and can be used in much the same way.

flavors pretty well and can be flavorful additions to flour when it will be used to dredge meat or fish before frying or baking. Mushroom powders can be used as flavorful thickeners, instead of flour or cornstarch, when you are making sauces and gravies. If you live in the Midwest and are saddled with hundreds of half-free morels, mushroom powders are a good way to utilize those giant stems. A cheap spice mill or coffee grinder works well for pulverizing dried mushrooms into powder.

An electric dehydrator is not absolutely essential for drying mushrooms. But it works best and doesn't cost much, and once you have one, you will find yourself using it throughout the year for drying herbs, fruits, and vegetables, making fruit leathers and jerky, and even for preserving your scientific mushroom specimens. (Yes, it's safe to dehydrate toxic species on the same dryer you use for food; just don't

Morels going into the dryer

Dried morels can make a very nice display.

Mushrooms should be dried until they are not pliable anymore. *Cracker-crisp* is the term used for a perfectly dried mushroom.

Dried mushrooms should be kept in an airtight container, out of direct sunlight. Glass jars, heat-sealed bags, and zip-top bags work well. They should be kept in a cool, dry place. A kitchen cupboard or pantry is ideal for long-term storage. Make sure to squeeze out as much air as you can from the bags without damaging the mushrooms inside. Jars should be filled as much as possible to limit the air space within. Jars can also be decorative pieces to place in the soffit space above kitchen cabinets.

Cooking with dried mushrooms is convenient. They can be added directly to a soup, or to a stock if you are making risotto. For dishes without a lot of liquid, the mushrooms can be rehydrated for a few minutes to an hour in a small bowl filled with water, stock, wine, or the like. Once they are rehydrated, remove the mushrooms by lifting them from the liquid. The leftover liquid will be flavorful and should be saved, but it could have some debris left in it. Pour it through a fine sieve and add to the dish you're preparing or save it for another use.

Canning and Pickling Mushrooms

Canning is an okay method for preserving an abundance of some cultivated mushrooms, but it's not really recommended for wild mushrooms; the end product is not much better than the canned mushrooms you can purchase at the store, and no one is wowed by them. The only safe way to can any food item that is not high in acid, salt, sugar, or some combination thereof is to use a pressure canner (cooker). This is easy and time efficient; follow the manufacturer's instructions. The old-fashioned boiling water bath method for canning is no longer recommend.

Pickling can be a suitable way to preserve some wild mushrooms, as the end result is a tasty and unusual snack that can be a nice addition to a spread of appetizers or served with cocktails. It is standard in Eastern European households to bring out pickled mushrooms when guests arrive. Furthermore, pickling can improve the taste of some strongly flavored mushrooms (e.g., honey mushrooms) and the texture of others (e.g., milk mushrooms).

For pickling, a pressure canner is not necessary, as the pickling brine's high amounts of acid, salt, and sugar inhibit bacterial growth, and the boiling temperature kills all molds and most other microbes present. Purchased pickling spice mixes work just fine, and they usually have a recipe on the package. The recipe typically consists of making a gallon of pickling brine from equal parts cider vinegar and water, to which you add 2–3 cups (396–594 g) sugar, 2–3 tablespoons (16–24 g) of salt, and an equal amount of pickling spice. (The authors do not like the flavor of cloves in their pickled mushrooms and recommend removing them if they are present.) Pack canning jars with mushrooms; slice them if you are using large ones and leave them whole if you are using very small ones. Add one clove of garlic to each jar. Mix all brine ingredients in a saucepan and bring to a boil. Pour the boiling liquid over the mushrooms in the jars. Boil the jars in a canner for 20 minutes. Let stand for one week before opening.

One final note: Jars of pickled (or dried) mushrooms make excellent gifts to fellow mycophiles!

Duxelles

Duxelles is a mixture of finely minced mushrooms that have been cooked in butter with other minced ingredients such as shallots, herbs, and wine, until the liquid from the mixture is mostly evaporated and the resulting paste is almost dry. Duxelles is an excellent addition to everything from eggs to ravioli filling, sauces, and stuffing placed inside meat prior to roasting. Simply spread on bread or toast, duxelles makes an excellent appetizer. It will keep for a while in the fridge, but it is best kept in small containers in the freezer.

Cooking with Mushrooms

When cooking mushrooms, the simplest techniques are most often the best. This is especially true with wild mushrooms. Simply sautéing and then adding them to any dish will turn something mundane into something special. Wild mushrooms can be used in fancy and complicated dishes, but the preparation of the mushroom itself still should be a simple sauté most of the time. Many chefs recommend the dry sauté method, rather than frying the mushrooms in butter. Dry sautéing is exactly what it sounds like: Place mushrooms in a medium-temperature skillet and cover. After about a minute, they will be getting hot, causing them to lose their water, which is essential to getting that nice caramelization that brings about savory flavors. Before they begin to lose water, you can hasten this with a small splash of water—just enough to make steam. Allow the steam to quickly cook the covered mushrooms, releasing water from them. Once the mushrooms have given off water, remove the lid and allow the water to evaporate. At the very end of cooking, add some butter, turn off the heat and move them around in the pan to coat. Add a bit of coarse salt and taste. Now you can decide what to do with that flavor and texture. Experimentation is the best way to find which mushroom will work best with each type of dish you are preparing, and personal tastes will vary.

When you are trying a wild mushroom for the first time, **this is important:** *Never mix any other mushrooms with a species that is new to you.* Although it may be a perfectly safe and edible species, you could have a sensitivity to it. Or you may not like the flavor or texture. A first-time mushroom should be prepared by the dry sauté method above. Try only a couple bites and reserve the remainder in the fridge for the next day. It's a good idea to keep one specimen uncooked, just in case.

Are there hazards when not cooking mushrooms—that is, eating them raw? In some cases, yes. Pretty much without exception, all mycological societies and mushroom clubs, as well as books about mushrooms, caution that you should not eat any wild mushroom without thoroughly cooking it first. This is mostly sound advice. Some mushrooms that are commonly eaten, such as *Agaricus* species, morels, and false morels, produce toxins that are removed or deactivated through cooking. Furthermore, mushrooms, like plants, are mostly made of indigestible fiber that is made easier to digest with cooking. And finally, if there are any microbial contaminants in your wild mushrooms, cooking may help to destroy them. Having said all this, there are some species of mushrooms—also like plants—that are best prepared raw, and we will include them in the pages that follow. Note that the amount of mushroom used is quite small, and you should moderate your intake if fibrous foods are normally problematic for you.

The culinary section of this book is mostly about simple techniques. In general, most mushrooms can be used interchangeably in any dish. If you have the good fortune to find a basket of edible mushrooms, or if you have a nice harvest of cultivated ones, try adding them to whatever dish you were already preparing. Mushrooms can often be substituted for meat in a dish. Like most mycophiles, we all have several (or several dozen!) mushroom cookbooks and recipes at home.

For this reason (and because our publishers scolded us about page count), we are more concerned with general techniques and ideas than with specific recipes. However, we do have a few special uses for certain mushrooms that we urge you to try in order to best enjoy them.

Broadly speaking, mushrooms in the kitchen can be put into two groups: Meaty and general-use mushrooms include most wild and cultivated edible mushrooms, such as morels, boletes, chanterelles, black trumpets, huitlacoche, oyster mushrooms, shiitakes, and others. Most preparations will benefit from adding any of these mushrooms or substituting them for meat. Aromatic or special-use mushrooms include candy caps and matsutake. These mushrooms can be used as others can, but much of their appeal comes from their enticing aromas, which may be overshadowed in most dishes. Furthermore, these mushrooms are expensive or otherwise hard to come by, so honor them with special treatment and it will be a memorable meal for you and your guests.

General Preparation

Cookbooks are typically laid out in a logical order that resembles the courses of a meal: breakfast, lunch, appetizers, soup, main course, and desserts. When cooking with wild mushrooms, however, you're most often preparing just a single dish such as a main course or sauce. For this reason, we begin with main courses and sauces.

Let's get started. As we mentioned, mushrooms make a great addition to many dishes and can be a substitute for meat. The easiest preparation is to simply slice, sauté, and add the mushrooms to your dish, along with any juices. My friend Albert Casciero, who frequently writes the culinary features for *FUNGI Magazine*, most often cooks mushrooms in the customary Italian mushroom style, *trifolatti*, which means sautéed gently with chopped parsley and garlic. This is similar to making duxelles, described previously. This preparation can be added to meat that has been pounded or sliced thin and rolled up, then roasted or grilled. Alternatively, you can

turn the mushroom preparation into something saucy by making a simple white roux with an added splash of wine at the end. Serving it over meat, rice, pasta, polenta, or toast wonderfully highlights the mushrooms.

Wild Mushroom Ragout

Chef Michael Blackwell is well known in Oregon, and he has made this sumptuous stew many times at the Breitenbush Mushroom Gathering. It is excellent served over pasta, rice, polenta, or simply thick slices of crusty bread, and alongside any roast meat. Use any meaty or general-purpose mushrooms—dried, fresh, or a combination.

Yield: 2 cups (450 g)

3 tablespoons (43 g) butter
2 tablespoons (30 ml) extra-virgin olive oil
1 cup (160 g) coarsely chopped onions
2 pounds (0.9 kg) fresh mushrooms, quartered or sliced
2 tablespoons (30 g) minced garlic
1 teaspoon (2 g) dried thyme, crushed
1 teaspoon (0.5 g) minced fresh rosemary
¼ cup (60 ml) dry sherry
1 cup (240 ml) heavy cream
2 tablespoons (30 ml) brandy
2 tablespoons (30 ml) fresh lemon juice
salt and freshly ground black pepper
pinch ground nutmeg

1. Heat butter and oil in a large, heavy pot over medium-high heat. Add onions and sauté until clear.

2. Add mushrooms and garlic and sauté 3–4 minutes, until tender.

3. Add herbs, sherry, and cream. Cook about 8 minutes, or until reduced and lightly thickened.

4. Add brandy and lemon juice. Cook another 2 minutes.

5. Season with salt, pepper, and nutmeg to taste. Serve.

Wild Mushroom Risotto

A serving of risotto is a common dinner course in Europe, especially Italy. Often it is the main course. Any of the meaty, general-purpose mushrooms will work in this dish, but there are two standouts: Far and away, risotto is best with black trumpets or king boletes. And seek out Arborio or Carnaroli rice, as they will take up much of the liquid during cooking and result in a much better risotto. Contrary to popular belief, making a good risotto is quite simple, though it does require two pots.

Yield: 4–6 servings

1 quart (946 ml) chicken or vegetable stock
About 2 handfuls dried or fresh mushrooms
Good-quality olive oil
A bit of diced onion or 1 shallot
1 cup (180 g) Arborio or Carnaroli rice
White wine
Parmesan cheese to taste

1. Bring stock, along with large pieces or slices of dried mushrooms (if using dried, unless they are crumbled dried black trumpets), to a slight boil in a stockpot.

2. Place some olive oil, diced shallot, and fresh chopped mushrooms (if using fresh) into a second pot or wide pan. Sauté over medium heat for a couple minutes.

Fresh black trumpets

3. Add rice and a few splashes of white wine and stir until absorbed. Add one ladle of stock to the pot (along with crumbled dried black trumpets, if using those) and stir until stock is absorbed. Add another ladle of hot stock. Continue this, one ladle at a time, stirring until absorbed each time, until the rice has puffed up and become tender, and all the stock is used up. Expect about 15 minutes total cooking time and don't rush it.

4. Stir in parmesan cheese and serve immediately. If risotto sits, it will continue to thicken; if this happens, add hot liquid to return it to creaminess just before serving.

5. If you have leftover risotto, save it for the following day. Make golf ball–sized balls of the cooled risotto, dip in egg wash, coat in breadcrumbs, and deep-fry to make a Sicilian treat, *arancini.*

Creamy Chicken Thighs with Morels

Dried mushrooms allow you to enjoy the flavors of mushroom season, even out of season. Without a doubt, the best mushroom dish that utilizes dried mushrooms was taught to me by my friend and writer Eugenia Bone. I have made it many times, and guests have always raved. And it is almost *too* simple.

For the preparation, you will notice that none of the ingredients have given amounts. That's because it depends on how many guests you are serving; figure one thigh per guest plus a couple leftover. (Some guests will eat two. Trust me.)

Serve this dish over pasta. Accompanied with a tossed salad of fresh lettuce and shaved fennel bulb, it will be a meal you will never forget.

Yield: varies

Butter
Shallot or two, finely diced
Chicken thighs, skin on
A handful or two of dried morels, depending on how much you love your guests (no need to rehydrate)
1 bottle sweet sherry
1 bottle semi-dry sherry
Heavy cream
Salt and freshly ground pepper

1. Select a deep pan that will accommodate all the chicken thighs in a single layer. Sauté shallot in butter. Once it's translucent, transfer to a plate.

2. Add a couple more tablespoons of butter to the pan and sauté chicken thighs on medium-high heat, skin side–down, in a single layer, for several minutes, until skin is very brown. Turn chicken pieces and continue cooking for several more minutes, until juices run clear.

3. Turn the heat down to medium-low and sprinkle cooked shallot over chicken.

4. Throw in morels. Add equal amounts of the sherries until chicken is covered. Cover and cook for 10–15 minutes, or until almost all the liquid is absorbed. You cannot overcook as long as there is still liquid in the pan.

6. Once cooked down, add cream until the sauce reaches your desired consistency. If you add too much, let it cook down.

7. Add salt and pepper to taste and serve over pasta.

Meatless "Chicken" with Sherry Cream Sauce

My friend Corrine is a mushroom hater. She will *not* touch them. But she loved this dish; even after she was stuffed, she continued to pick the tender lemony pieces of "chicken" out of the leftovers, relishing them. I told her it was actually mushrooms . . . the next day. This is an all-time classic dish, and you can use any of the meaty, general-purpose mushrooms, but it is the absolute best dish for chicken of the woods (*Laetiporus* species), which has a taste and texture similar to chicken breast.

And the dish could not be simpler to make. It is excellent served over pasta, rice, polenta, or simply thick slices of crusty bread. No one will suspect it's meatless.

Yield: 3 cups (600 g)

3 tablespoons (42 g) butter or (44 ml) olive oi
1 coarsely chopped onion or 2 shallots
1 carrot, chopped
1 stalk celery, chopped
1 pound (454 g) fresh mushrooms, ideally chicken of the woods, broken or sliced
2 tablespoons (15 g) flour
2 cups (480 ml) chicken or vegetable stock
1 teaspoon fresh herbs of your choice (chives, thyme, marjoram, oregano, and tarragon are all nice)
¼ cup (60 ml) dry sherry
2 tablespoons (30 ml) fresh lemon juice (or the juice of a half lemon)
¼ cup (60 ml) heavy cream
salt and freshly ground black pepper to taste

1. Sauté onion, carrot, and celery in butter over medium heat until becoming tender.

2. Add mushrooms and cover, cooking for several minutes, until liquid comes out of the mushrooms.

3. Add flour and stir, making a roux. Slowly add stock and stir continuously until thickened.

4. Add herbs and sherry. Simmer for a couple minutes.

5. Just before serving, add lemon juice and enough cream for the mixture to reach your desired thickness. Add salt and fresh ground pepper and serve.

Matsutake Gohan Rice

My dear friend Patrice Benson was a wonderful mycophile and mushroom cook. She lived in Seattle and enjoyed cooking matsutake mushrooms, which can be abundant there. Patrice introduced me to the most famous dish prepared with this mushroom. Matsutake Gohan, or Pine Mushroom Rice, results in a fragrant rice dish that exemplifies the essence of Japanese cuisine. The simplicity of this dish belies the depth of complex flavors that come together in this humble bowl of rice, and its presence on menus heralds the coming of fall.

Yield: 6 cups (900 g)

2 cups (400 g) Japanese short grain rice
1 ½ cups (360 ml) water or kombu dashi (stock made from seaweed)
1 teaspoon (5 ml) soy sauce
1 teaspoon (5 ml) sake
4 ounces (113 g) matsutake mushroom (one large mushroom or two small)
Salt

1. Place first four ingredients into a heavy pot with a tight-fitting lid, or a rice cooker.

2. Slice matsutake and place on top of rice and cook in a rice cooker. Or in a pot on the stove, turn heat down to low once liquid comes to a boil. The mushroom will steam, flavoring the rice, but will retain a firm texture. Be sure to take the rice off the heat once all the liquid has been absorbed.

3. Salt to taste and serve.

Matsutake Grilled in Foil

My friend Milton Tam also lives in Seattle and is also a member of the Puget Sound Mycological Society there. He introduced me to this method of cooking matsutake mushrooms, and it's my favorite. You can experiment with adding a drizzle of sesame oil, mirin, sesame seeds, or whatever else along with the soy sauce. Serve the resulting mushroom juice over rice or in a small serving dish to use for dipping. It will have an amazing flavor.

Yield: 1 serving

Matsutake mushroom
A couple pats butter
Drizzle light soy sauce

1. Slice matsutake mushroom lengthwise and place slices in foil. Add butter and soy sauce.

2. Tightly seal into a foil pouch and place on a hot grill, or right onto the medium flame of a gas stove.

3. Once you hear sizzling inside the pouch, allow mushroom to steam in its own juices for a couple minutes. Take off the heat. Allow to cool for a minute or two.

4. Plate mushrooms and pour the juice over rice or into a small serving dish to use for dipping. The mushrooms should be nicely cooked but still firm and toothsome.

Matsutakes

Grilling Mushrooms

Grilling mushrooms is a no-brainer. Big portobellas and shiitakes are showstoppers. But some wild mushrooms are particularly well-suited to grilling, namely matsutake and hen of the woods.

Yield: varies

Mushrooms of your choice
Olive oil or sesame oil
Soy sauce to taste

1. Brush mushrooms with oil and soy sauce.

2. Grill, keeping the heat at medium or lower, and indirect to prevent drying out, until mushrooms are fully cooked.

3. Serve with or without a dipping sauce.

Portobellas and Shiitakes: Can be used whole, placed gills up. Just spoon a little butter or oil, soy sauce, and minced garlic right onto the gills.

Matsutake: Can be sliced lengthwise, thickly, and grilled with intermittent basting.

Hen of the Woods: Use entire pieces pulled or sliced from the entire fruit body, and grill with basting.

Large Morels: For something really amazing, try the method my friend Denis Benjamin introduced me to a few years ago. We'd had a successful day of morel picking near my home in Wisconsin, and he proposed we grill the largest of the morels. He placed them whole, right on the grill, and lovingly basted them continuously with butter. He never let them out of his sight, being

careful to turn each one a quarter of a turn or so, every minute or two. Once perfectly cooked and sprinkled with some coarse salt, they were absolutely decadent.

Deep-frying Mushrooms

We've all had pretty mediocre deep-fried mushrooms at carnivals and fairs. Between the tired, worn-out cooking oil and the gloppy, thick batter, it's no surprise that they are so unsatisfying. But fried mushrooms can be very good. During king bolete season in Italy, they're on most restaurant menus. Thick slices and dredging in a little seasoned flour or corn meal is all that is required. Fried until golden brown, sprinkled with coarse salt, and sprinkled with a squeeze of lemon is as good as a mushroom gets—crispy on the outside and juicy inside.

Even better are tempura-battered fried mushrooms. A tempura batter is light and watery, and when fried it becomes very crispy, like ultra-thin glass. (To make tempura batter, whisk a cup [240 ml] of club soda with a cup [160 g] of rice flour. Wheat is a decent

substitute. The batter should be very watery.) For this method, mushrooms with a lot of structure to hold in little bits of thin batter work best. Try morels sliced in half. Or try lion's mane and other *Hericium* species—this is a favorite way to prepare them. Simply pull apart little bite-sized pieces of lion's mane and dunk them into the tempura batter. Lift out and allow nearly all the batter to drip off. Then drop the pieces, one at a time (so they don't clump together) into hot oil. Do in small batches so they are not crowded. After a couple minutes, when they are golden brown and very crispy, remove them with a spider or slotted spoon and transfer them to a paper towel. Serve them with a Thai-style or other dipping sauce. All those little spines will be wonderfully crunchy on the outside, each nugget tender and juicy inside. Tempura-battered mushrooms are the gateway to getting kids to like mushrooms!

Mushroom Quiche

This is another no-brainer that can be served any time of the day, and it can be made with ingredients you likely have on hand. Homemade pie crusts are fantastic, but in a pinch, good quality store-bought crusts are a great time saver and last in the freezer for months. They work best if left to thaw in the fridge overnight. Any of the meaty, general-purpose mushrooms will work fine for this one. I especially enjoy fresh blewits (*Clitocybe nuda*) prepared this way. And there is no better cheese to go with eggs than Gruyère. Other cooked vegetables and meats, including sliced leeks, broccoli, and ham, make excellent additions to the mixture before baking.

Morels with asparagus and wild ramps

Blewits

Yield: 1 quiche

Prepared pastry crust

1 tablespoon (14 g) butter

1 shallot, finely minced

1 pound (454 g) fresh mushrooms, coarsely chopped, or 1 ounce (28 g) dried and rehydrated

3 eggs, beaten

1 cup (240 ml) whipping cream

¼ pound (113 g) Gruyère cheese, grated

Salt and freshly ground pepper to taste

1. Place pastry crust in a pie plate (pie beads are helpful to keep the crust's shape). Bake it at 350°F (177°C, or gas mark 4) for about 15 minutes, until brown.

2. Meanwhile, sauté shallots in frying pan with butter over medium heat until translucent.

3. Add mushrooms and continue cooking until mushrooms are tender and liquid has evaporated.

4. Spread cooked mushrooms over cooked pastry crust.

5. In a medium-sized bowl, mix eggs, cream, and cheese. Pour over mushrooms in pastry shell.

6. Bake for 35 minutes, until center is set and top is slightly brown.

Tavis's Savory Mushroom Tart

Store-bought puff pastry is wonderfully easy to work with, a major time saver, and versatile with mushroom cookery. This recipe is similar to a strudel, but the prepared mushrooms are simply served over the crispy pastry. The amounts of below are for fresh herbs; use a large pinch of each for dried herbs.

Yield: 2 servings

1 sheet puff pastry
½ cup (100 g) chopped Shiitake, stems removed
½ cup (100 g) chopped Oyster mushrooms
1 garlic clove, minced
Salt and freshly ground pepper to taste
⅓ cup (80 ml) demi-glaze or beef stock
3 tablespoons (44 ml) cream
1 tablespoon (4 g) chopped parsley
1 tablespoon (2 g) rosemary
1 tablespoon (2 g) thyme
Goat cheese to taste

1. Preheat oven to 350°F (177°C, or gas mark 4).

2. Cut puff pastry to desired shape. Bake for 10 minutes.

3. Sauté mushrooms at medium-high heat until hard-seared. Add garlic, salt, and pepper. Add demi-glaze, cream, and herbs.

4. Simmer down by half and serve over puff pastry. Add goat cheese and serve.

Mushroom "Caviar" and Mushroom Cheese Spread

Both of these recipes come from the original "Amanita Rita," Rita Rosenberg, a chef featured at the Telluride Mushroom Festival since the early days of the festival. Her wild mushroom strudels are legendary. But these two appetizer dips absolutely must be tried, especially the caviar. Any mushrooms whatsoever will work nicely for either preparation, but I prefer to use a mix of dried morels and dried black trumpets that have been rehydrated. To begin, start by making simple duxelles or *trifolatti* (see page 132). And do not throw out the water after rehydrating mushrooms. Pass it through a fine sieve and add to the duxelles as it cooks down. Once the liquid evaporates, its flavor will make the mushrooms extra intense.

Yield: 1 cup (240 g)

CAVIAR INGREDIENTS
1 cup (240 ml) duxelles
Juice of half a lemon
1 tablespoon (15 ml) Worcestershire sauce
2 tablespoons (30 ml) good mayonnaise
Salt and ground black pepper to taste

1. Put duxelles, lemon juice, Worcestershire sauce, and mayonnaise into a blender.

2. Pulse the blender a few times. You want the mixture to blend but retain a coarse, caviar-like texture.

3. Salt and pepper and serve with crackers or toast points.

Dried black trumpets, candy caps, and morels

Yield: 2 cups (425 g)

CHEESE SPREAD INGREDIENTS

1 cup (240 ml) duxelles
8 ounces (227 g) cream cheese, room temperature
1 tablespoon (14 g) butter, room temperature
Salt and ground black pepper to taste

1. Put duxelles into a food processor. Add cream cheese and butter. Blend until smooth.

2. Salt and pepper and serve with crackers or toast points.

Mushroom Pâté

Here is another recipe from Chef Michael Blackwell of Oregon that has been a huge hit at the Breitenbush Mushroom Gathering. It is easy to make and delicious, and it works well with dried mushrooms, so you can make it year-round. It's best served right out of the dish and spread on crusty bread, but it's also great as a sandwich spread or filling for dumplings, or for an elegant presentation.

Yield: about 1 pound paté, enough to serve 6 or more

1½ pounds (680 g) white or brown mushrooms from grocer, thinly sliced

1 ounce (28 g) dried mushrooms (such as porcini, boletes, or morels)

½ cup (120 ml) hot water, for rehydrating dried mushrooms

4 tablespoons (56 g) butter

2 bunches green onions, finely chopped

1 pound (454 g) chicken livers

½ cup (120 ml) sherry, Madeira, Marsala, or brandy

½ teaspoon (2.5 ml) hot sauce

½ teaspoon (2 g) granulated garlic

¼ teaspoon ground white pepper

¼ teaspoon ground nutmeg

⅛ teaspoon ground cloves or allspice

½ cup (120 ml) heavy cream

1. Rehydrate dried mushrooms in the hot water. Place butter in pan and sauté all mushrooms (including rehydrated ones) over medium heat until they begin to soften.

2. Add green onions and sauté 1 more minute.

3. Make a window in the center of the pan and add livers.

4. Pour wine around outside edge of pan. Let cook a few minutes.

5. Turn livers to cook other side. Livers are done when firm but still pink in center. Remove livers from pan and set aside.

6. Add remaining ingredients to liquid in pan and simmer until reduced to less than ¼ cup (60 ml).

7. Remove from heat and transfer to the bowl of a food processor, along with reserved livers. Using a steel blade, process until mixture is completely smooth. Be patient.

8. When mixture is smooth, transfer it to wide-mouthed short jars, molds, or ramekins. Let cool in refrigerator. This will keep nicely for several days in the fridge if unopened. Once opened, it should be used up within a day or two.

Double Oyster Chowder

This is without doubt the best mushroom soup. It came to me by way of Joe McFarland, a naturalist and lover of mushrooms and food in southern Illinois. You don't have to use oysters; you can substitute chunks of fish, shrimp, chicken and corn, or something else. You don't have to use oyster mushrooms, either. But there is magic that happens when you get oysters and oyster mushrooms together!

Yield: 8 servings

5 tablespoons (70 g) butter

1 teaspoon (5 g) minced garlic

3 cups (702 g) chopped oyster mushrooms

3 tablespoons (23 g) flour

1 cup (240 g) dry white wine

1 quart (946 ml) chicken stock

2 cups (480 ml) heavy cream

Salt and freshly ground black pepper

3 oysters per person, optional

Chopped parsley for garnish

1. Sauté garlic and mushrooms in 2 tablespoons (28 g) butter over medium heat until they are becoming tender.

2. Add remaining butter and flour to make a roux.

3. Add wine slowly to liquify roux; it will thicken quickly and as it does, add chicken stock, 1 cup (240 ml) water, and cream. Season with salt and pepper.

4. Just before serving, add oysters. They only need to cook for about 2 minutes.

5. Serve by ladling into bowls and garnishing with parsley.

Spicy Oyster Mushroom Soup

Most cultivated mushrooms will probably work with this one. To make it a hot and sour–style soup, just add a few big glugs of rice wine vinegar before serving.

Yield: 16 servings

1 onion, diced
2 or 3 carrots, diced
2 or 3 celery stalks, diced
2 pounds (0.9 kg) oyster mushrooms, sliced
1 tablespoon (14 g) minced garlic
1 tablespoon (14 g) minced ginger
4 quarts (3.8 L) chicken or vegetable stock
1 ounce (30 ml) hot sauce, or to taste
3 ounces (90 ml) soy sauce
⅔ cup (85 g) cornstarch

1. Sauté onion, carrots, and celery in a large pot until onion is translucent.

2. Add mushrooms, garlic, and ginger. Cook until mushrooms are fully cooked.

3. Add stock, hot sauce, and soy sauce and bring to a boil.

4. Make slurry with corn starch and 1 cup (240 ml) water and add it to boiling soup. Stir together and serve.

Lobster Mushroom Consommé

If you are in the mood for a much lighter soup, it doesn't get much better than this one, which I learned from Albert Casciero. And it's the best way that I know to highlight the taste and

texture of lobster mushrooms. Here, the striking color of the mushroom lends an interesting golden hue to the broth in contrast with the deep red of its surface. Dashi is a Japanese powdered stock base usually flavored by dried bonito fish, kelp, or the like. It's widely available in food markets, it's invaluable for making miso soups, and it lasts a long time on the shelf.

Yield: varies

Lobster mushrooms, cleaned and sliced
2 tablespoons (28 g) butter
Dashi
A few strips of dried seaweed
Fresh snipped chives or spring onion for garnish

1. Sauté slices of lobster mushrooms in butter.

2. Make a broth with water and dashi.

3. Add mushrooms to broth. Seaweed will further enhance the look and flavor of this serving.

4. Garnish with chives or spring onion and serve.

Mushroom Crêpes

This recipe comes from Albert Casciero. Although he is Italian, he can cook French food—including crêpes—masterfully. Crêpes are easy to make, although they require a bit of initial dexterity. Once mastered, they are a wonderful element of many meals, both savory and sweet. And it's not essential to purchase special equipment, but that equipment is useful and inexpensive. Once you have three key pieces, you will likely use them all the time, as we do at our home. A skillet will work, but a flat griddle or crêpe pan works much better; they come in electric or cast iron, which sits over a stovetop flame. A special crêpe spatula is also handy, and the third piece of equipment is the spreader for covering the cooking surface with an even layer of batter. If you are unsure about any crêpe technique or equipment, you will find numerous videos online. The great thing about crêpes is that anything goes, including leftovers. Some sautéed mushrooms, cooked meat such as ham or chicken (leftovers work fine), and cheese (Gruyère is our favorite) is an excellent combination. Or you can go the sweet route and top your crêpes with jelly, jam, or Nutella. We always do both dinner and dessert crêpes. Kids love them and love making them their way!

Chanterelles

Smooth chanterelles, *Cantharellus lateritius*

Hedgehog mushrooms, *Hydnum repandum*

Yield: About 12 crêpes

1 cup (128 g) wheat flour
2¼ cups (540 ml) milk
2 eggs
Butter for pan
Toppings of your choice

1. Whisk flour and milk in a large bowl. Add eggs and mix well with the whisk. You want a very watery liquid mixture. For very thin and light crêpes, you may want to add even more milk.

2. Set your cooking surface to medium-high heat. When surface is hot, lightly grease it, for example by getting a little butter on a wad of paper towel and quickly wiping the entire cooking surface.

3. Scoop about ¼ cup (60 ml) batter with a ladle and pour it onto the pan. Spread batter evenly with batter spreader. Alternatively, you can spread batter by tilting and rotating the pan, but it's not as easy.

4. After about a minute, when the top of crêpe no longer appears wet, flip it. After just a few seconds, it will be ready to serve. The crêpe

should be golden brown and pliable but not crispy.

5. Once your crêpe comes off the stove hot, immediately top with cheese, if using, and then other ingredients. Roll up and serve.

Porcino Crudo Salad

Each autumn during king bolete season, I witness a form of madness in Italy. Italians are crazy about king boletes, and I must confess that I, too, have the madness. During porcini season, many restaurants serve dishes featuring this mushroom. The best specimens are not cooked—they are reserved for this raw salad. It is simple, but it is the best way to experience the full aroma and flavor of this mushroom. Yes, the mushrooms are raw, but usually shaved very thin using a truffle shaver, so a serving contains very little mushroom. If you don't own a truffle shaver, you should get one. A cheap one from the internet is only about five bucks, and you can also use it to create fancy thin curls of parmesan to top pasta or chocolate to top a dessert. A truffle shaver is simple to use but keep your finger away from the razor-sharp blade.

Simple shaved raw king boletes with lemon, olive oil, and salt.

Porcini crudo salad

This is wild mushroom cuisine at its best. I have never seen anyone try this without swooning.

Yield: varies

Torn lettuce
Thin slices carrot, fennel bulb, or other salad vegetables
Small, very firm king bolete mushroom
High-quality olive oil
Lemon juice
Coarse salt

1. Create a bed of torn lettuce and slices of salad vegetables.

2. Shave king bolete over the entire salad.

3. Drizzle with olive oil and a generous squeeze of lemon juice. Sprinkle with coarse salt. Serve.

Giant Puffball S'mores

Mushrooms can be turned into all sorts of creative snacks and finger foods. How about something really unusual? This is the best way I know of to prepare giant puffballs. The recipe was given to me by Joe McFarland, and it's never failed to impress.

Yield varies

Bacon
Giant puffball mushroom, sliced ⅓ inch (8 mm) thick
White bread
2 tablespoons (28 g) butter
Salt
Swiss or Gruyère cheese
Chives

1. Preheat oven to 400°F (200°C, or gas mark 6). Cook bacon until nearly crisp by your preferred method. (I like to do mine on a cookie sheet in the 400°F oven.)

2. Using a round cookie cutter, cut an equal number of disks from mushroom and bread slices.

3. Heat butter in medium-high skillet and sauté mushrooms until golden brown. Salt to taste.

4. Assemble the s'mores on a cookie sheet. (I like to reuse the same one I used for bacon; simply scrape off most of the bacon fat first.) Stack a mushroom slice on top of a round of bread, and then top with bacon, cheese, and chives in that order.

5. Bake 5–7 minutes, or until cheese melts. Serve.

Mushroom Sauces and Gravy

Every cook has a "secret sauce" for improving a dish. The best sauces are the ones that are easy to make and versatile, going nicely with vegetables, meats, fish, and even crusty bread. A béchemel, or simple white sauce, can get kids excited about leftovers. A mornay sauce is more exciting, but nothing more than béchemel with the addition of parmesan and a splash each of Worcestershire and sherry.

Bordelaise is a fancy name for a basic mushroom sauce made by starting with a simple duxelles or *trifolatti* (see page 152), adding a cup (240 ml) each of red wine and brown bone stock, and letting the whole thing cook down to the desired thickness. If starting with dried mushrooms, do not throw out the water after rehydrating them. Pass it through a fine sieve and add it to the cooking sauce.

A morel cream sauce is just about everyone's favorite wild mushroom sauce. The easy part is making it: Just sauté sliced or chopped morels in butter for a few minutes until thoroughly cooked. Add a tablespoon or two (8–15 g) flour and mix to make a roux. Slowly add heavy cream until the sauce reaches your desired thickness.

Morel cream sauce over a pork chop

Do not allow this to become super thick. It's much nicer and doesn't cover up (literally) the meat or vegetables if the morel sauce is kept nice and thin. If it gets too thick, simply thin it with some white wine, dry sherry, Madeira, brandy, or the like. I've made this countless times and used just about any kind of wine you can think of. Freshly cracked peppercorns or whole green peppercorns are a nice addition.

My friend Tom Nauman in northern Illinois takes morel cream sauce one step further. Tom is famous in morel circles as he has run the Illinois State Championship Morel Hunt for many years, and he has been featured in numerous books, newspapers, and television programs. His biscuits and morel gravy recipe

is legendary in the Midwest. Having stayed at his home, I have witnessed Tom and his wife Vicky preparing this dish, and I know the secrets of the sauce. They fry up batch after batch of morels, simply in butter in a big skillet. The key is in not cleaning it between batches. You can do the same. When you're ready to make Tom's intense morel gravy—and it could be the next morning—begin by frying coarsely chopped breakfast sausage in that same well-seasoned skillet. When it is browned, plate the sausage and set it aside. Next, add butter and lots of chopped morels to the hot skillet. Once they are adequately cooked, add a couple tablespoons (15 g) flour and stir to make a roux. Then slowly begin adding milk, stirring constantly to prevent clumps. Add milk until sauce reaches the desired thickness. Add the sausage and its juices back to it. Serve over biscuits or toast.

Candy Cap Mushroom Cookies

I think I first heard about this recipe from one of David Arora's wonderful books on West Coast mushrooms. The candy cap mushroom is an unusual milk mushroom. When you encounter it in the woods, it can be hard to distinguish from the myriad other small *Lactarius* species. All descriptions note the peculiar odor of maple syrup, but when it's fresh, the mushroom typically has no scent whatsoever. Older specimens may have a slight aroma. However, as

the mushrooms dry, the odor intensifies—and it seems to get stronger and stronger as they age. I have collections that have been double- and triple-bagged and still the entire cabinet reeks of candy caps! Because of their sweet smell, candy caps are most often used in desserts. They actually have no real flavor, but the fragrance in your mouth convinces you otherwise.

Yield: about 6 dozen cookies

1 cup (about ¼ ounce, or 7 g) dried candy cap
    mushrooms
1 cup (224 g) butter, softened
1 cup (192 g) sugar
1 egg
2½ cups (300 g) flour

Two species of candy cap mushrooms, California

1. Crumble dried candy caps with your fingers until they are fine but not quite powdered.

2. Cream together butter and sugar in a mixer. Beat in egg.

3. Add flour and mix. Add crumbled mushrooms and mix very well.

4. Form dough into a long roll, about 1–1 ½ inches (2.5–3.8 cm) thick. Wrap it in wax paper and chill for a few hours or overnight.

5. Preheat oven to 350°F (177°C, or gas mark 4). Slice rolled dough into cookies and place them on ungreased cookie sheets, spaced well apart. Bake for 8–10 minutes, or until lightly browned at edges, and then let cool on a wire rack.

Huitlacoche Ice Cream

I know what you're thinking. Huitlacoche *does* look weird! And although I'm certain my parents would refuse to eat it to this day, I assure you it is delicious. No one I've ever served it to for the first time has ever had anything but high praise for it. It has an amazing, understated blend of mushroomy-sweet corn flavor and hints of chocolate. Even more amazing is how well the flavors go with each other and with all sorts of savory dishes.

It's easy to overpower huitlacoche, so if you are cooking some up, take it easy on strong flavors, including garlic. In Mexico, it's typically lightly sautéed with onion (I prefer shallot), garlic, chiles (poblanos are best and do not overpower it), and cilantro. And that's about it. When huitlacoche is in season, this mixture is then added to soups, sauces, and casseroles; stuffed into any meat and roasted; or used as a filling for quesadillas and tacos. I've cooked a lot of huitlacoche and enjoy it worked into just about anything.

Fresh huitlacoche in the kitchen

But my all-time favorite preparation, one that has impressed many houseguests, is huitlacoche ice cream. Yes, ice cream! Mushroom mycophagy events often struggle to serve a mushroom-based dessert and usually end up with lackluster results. But this is one sweet mushroom dish that really will impress, and I hope you give it a try. Fresh or frozen huitlacoche will work equally well. I do not recommend canned, as it looks gross and has no flavor.

Yield: 1 quart of ice cream, about 8 servings

6 ounces (170 g) huitlacoche, fresh-picked or frozen
2 tablespoons (28 g) butter
¼ teaspoon salt
1 cup (240 ml) milk
3 cups (720 ml) heavy cream
¾ cup (180 ml) egg yolk (from 9–12 eggs)
9 ounces (255 g) sugar

1. Cut all huitlacoche, if it is still on the cob, along with any corn kernels, using a sharp knife.

2. Melt butter in a heavy saucepan. Add huitlacoche (and corn), and cook over medium heat for just a few minutes, until the mushroom pieces look shiny.

3. Add salt and place cooked huitlacoche in a blender. Add enough milk to purée, and purée.

4. Return puréed mushroom mixture to the same saucepan. Add remaining milk and heavy cream and bring to a boil. Reduce heat.

5. Whisk egg yolks and sugar in a separate bowl until ribbony. Fold into saucepan. Let mixture simmer until it thickens, but do not boil.

6. Cool thickened mixture over ice or in the freezer until very cold but not quite freezing. Pour mixture into an ice cream machine.

Resources

List of Suppliers

Field and Forest Products
Peshtigo, WI
fieldforest.net
800-792-6220

Fungi Ally
Hadley, MA
fungially.com
978-844-1811

Fungi Perfecti
Olympia, WA
fungi.com
800-780-9126

Midwest Grow Kits
Algonquin, IL
midwestgrowkits.com
800-921-4717

Mushroom Mountain
Easley, CA
mushroommountain.com
864-855-2469

Mushroom People
Summertown, TN
mushroompeople.com
931-964-4400

Smugtown Mushrooms
Rochester, NY
smugtownmushrooms.com
585-690-1926

Stockport Gourmet Mushrooms
Heath, OH
740-323-0793

Superior Mushrooms, LLC
Mason, WI
regan@superiormushrooms.com
715-413-1341

Sylvan Inc.
Seventeen locations serving more than
65 countries
sylvaninc.com

Where to Get Even More Information

Australasian Mycological Society
australasianmycologicalsociety.com

Beginning Farmers
beginningfarmers.org

European mushroom information
fungus.org.uk

European Mycological Association
euromould.org

Fungal Network of New Zealand
New Zealand Mycological Society
funnz.org.nz

FUNGI Magazine
www.fungimag.com

The Fungi of California
mykoweb.com/CAF

Mushroom Expert
mushroomexpert.com

The Mushroom Growers' Newsletter
mushroomcompany.com

North American Mycological Association
namyco.org

Northeast Mycological Foundation
newf.org

ShiiGA (Shiitake Growers Association)
shiigaw.org

About the Authors

Britt Bunyard, Ph.D., is the founder, publisher, and editor-in-chief of the mycology journal *Fungi*. Britt has worked academically (and played very amateurishly) as a mycologist his entire career, writing scientifically for many research journals and popular science magazines. He has served as an editor for mycological and entomological research journals, and mushroom guide books. A popular evangelizer on all things fungal, Britt has been featured on NPR's *All Things Considered*, PBS's *NOVA* and *Wisconsin Foodie* television programs, and in *The Atlantic*, *Vogue*, *Forbes*, *Saveur*, *Women's World*, and others. He serves as Executive Director of the Telluride Mushroom Festival. Bunyard co-authored *Mushrooms and Macrofungi of Ohio and Midwestern States* (The Ohio State University Press) and *Amanitas of North America* (The FUNGI Press).

Tavis Lynch is an amateur mycologist from northern Wisconsin. He has been studying wild fungi since he was a child and has been a cultivator of mushrooms for the past two decades. He teaches several mycology classes in 22 schools in the Upper Midwest. Tavis has made numerous appearances on Wisconsin Public Radio and the occasional television appearance to discuss wild fungi and the cultivation of gourmet mushrooms. He is the author of *Mushroom Cultivation* (Quarry Books).

Acknowledgments

Tavis: I would like to thank my Aunt Jill for being my hero.

Britt: Many thanks to the dozens of mycological societies and mycophiles from across North America (and beyond), who have hosted me over the years (and even taken me to their secret spots), thus allowing my study and photography of so many of the species depicted in this book. This book would not have been possible without your help!

Index

Photo credits